David G. Costa

An Invitation to Variational Methods in Differential Equations

Birkhäuser
Boston • Basel • Berlin

David G. Costa
University of Nevada, Las Vegas
Department of Mathematical Sciences
4505 Maryland Parkway
Las Vegas, NV 89154-4020
U.S.A.

Cover design by Alex Gerasev, Revere, MA.

Mathematics Subject Classification: 35J20, 35J25, 35J60, 58E05, 34L30, 55M05

Library of Congress Control Number: 2007925919

ISBN-13: 978-0-8176-4535-9 e-ISBN-13: 978-0-8176-4536-6

Printed on acid-free paper.

9 8 7 6 5 4 3 2 1

www.birkhauser.com (TXQ/EB)

To my wife Gabriele, for encouraging me

to write this little book

Contents

Preface

This little book is a revised and expanded version of one I wrote for the "VIII Latin-American School of Mathematics" [29], in Portuguese, based on which I have periodically taught a topics course over the last 18 years. As the title suggests, it is an introductory text. It is addressed to graduate students of mathematics in the area of differential equations/nonlinear analysis and to mathematicians in other areas who would like to have a first exposure to so-called variational methods and their applications to PDEs and ODEs. Afterwards, the reader can choose from some excellent and more comprehensive texts, which already exist in the literature but require somewhat more maturity in the area.

We present a cross-section of the area of variational methods, with a minimum (no "pun" intended) of material, but clearly illustrating through one or two examples each of the results that we have chosen to present. So, besides the first motivating chapter and an appendix, there are only ten short chapters (with three or, at most, four sections each) through which the reader is quickly exposed to a few basic aspects of the beautiful area of variational methods and applications to differential equations. In fact, the reader may initially skip some of the more technical proofs of the main theorems, concentrating instead on the applications that are given.

Chapter 1 is of a motivating nature where we present five simple ODE problems which, from a variational point of view, illustrate existence results in situations of "minima" and "minimax," as well as a nonexistence result in a situation of resonance. In fact, to whet the reader's appetite, we present through some sketches a sneak preview of the geometry of each of the five functionals associated with the given problems. This chapter sets the stage for the topics to be covered in the following chapters: minimization, deformation results, the mountain-pass theorem, the saddle-point theorem, critical points under constraints, a duality principle, critical points in the presence of symmetries, and problems in which there is lack of compactness. At the end of Chapters 2 through 6 we provide a few exercises to the reader.

Of course there were a number of other important topics that were not covered in this little book. Our main goal, we emphasize, was to offer the reader a sampling of topics in critical point theory serving as a stepping stone to a number of excellent and comprehensive texts which exist in the literature, such as [63] (1986) by P. Rabinowitz, [55] (1989) by J. Mawhin and M. Willem, [69] (1990) by M. Struwe, and [75] (1996) by M. Willem.

David G. Costa
University of Nevada Las Vegas
Fall 2006

Some Notations and Conventions

- $\mathbb{R}$, $\mathbb{Z}$ and $\mathbb{N}$ denote the set of real numbers, integers and nonnegative integers.
- $\mathbb{R}^N$ denotes the usual N-dimensional Euclidean space.
- $\Omega \subset \mathbb{R}^N$ denotes an open set (usually with a smooth boundary) $\partial\Omega$.
- $|E|$ denotes the Lebesgue measure of a measurable set $E \subset \mathbb{R}^N$.
- χ_E is the characteristic function of the set E.
- $L^p(\Omega)$, $1 \leq p < \infty$, denotes the space of measurable functions u on Ω with norm $\| u\|_{L^p} := \left(\int_\Omega |u|^p\, dx\right)^{1/p} < \infty$.
- $L^\infty(\Omega)$ denotes the space of measurable functions u with $|u(x)| \leq C$ a.e. in Ω with norm $\| u\|_{L^\infty} := \inf\{C \geq 0 \mid |u(x)| \leq C \ \textit{a.e. in}\ \Omega\}$.
- $\| \cdot \|_X$ (or simply $\| \cdot \|$) denotes the norm in the space X.
- $\langle \cdot , \cdot \rangle_X$ [or $(\cdot , \cdot)_X$] denotes the inner-product in X.
- $C_0^\infty(\Omega)$ [resp. $C_0^\infty(\mathbb{R}^N)$] denotes the space of infinitely differentiable functions with compact support in Ω [resp. $\mathbb{R}^N$].
- $C^k(\overline{\Omega})$ denotes the space of k-times continuously differentiable functions on $\overline{\Omega}$ [with $\partial\Omega$ assumed smooth].
- Δu denotes the Laplacian of u, $\sum_{i=1}^N u_{x_i x_i}$.
- $C^1(X, \mathbb{R})$ is the space of continuously differentiable functionals on X.
- $H_0^1(\Omega)$ [resp. $H^1(\Omega)$] is the Sobolev space obtained by completion of $C_0^\infty(\Omega)$ [resp. $C^\infty(\overline{\Omega})$] in the norm $\|u\| := \left(\int_\Omega [|\nabla u|^2 + |u|^2]\, dx\right)^{1/2}$.
- $W_0^{1,p}(\Omega)$ [resp. $W^{1,p}(\Omega)$] is the Sobolev space obtained by completion of $C_0^\infty(\Omega)$ [resp. $C^\infty(\overline{\Omega})$] in the norm $\|u\| := \left(\int_\Omega [|\nabla u|^p + |u|^p]\, dx\right)^{1/p}$, $1 \leq p < \infty$.
- $H^1(\mathbb{R}^N)$ [resp. $W^{1,p}(\mathbb{R}^N)$] is the Sobolev space obtained by completion of $C_0^\infty(\mathbb{R}^N)$ in the norm $\|u\| := \left(\int_\Omega [|\nabla u|^2 + |u|^2]\, dx\right)^{1/2}$ [resp. $\|u\| := \left(\int_{\mathbb{R}^N} [|\nabla u|^p + |u|^p]\, dx\right)^{1/p}$], $1 \leq p < \infty$.

- $\operatorname{Lip}_{\text{loc}}(X)$ denotes the space of locally Lipschitzian functions on X.
- (u_n) denotes a sequence of functions.
- $u^+ := \max\{u, 0\}$ (resp. $u^- := \max\{-u, 0\}$) denotes the positive (resp. negative) part of u, so that $u = u^+ - u^-$.
- The arrow $\rightharpoonup$ (resp. $\rightarrow$) denotes weak (resp. strong) convergence.
- **Note to the Reader**: In each chapter, all formulas and statements are denoted and referred to by a simple double index $x.y$, where x is the section number. However, when referring to formulas and statements from another chapter, we use a triple index $z.x.y$, with z indicating the chapter in question. In addition, we shall write Section $z.x$ when referring to Section x of Chapter z.

1

Introduction

It is fortunate (for some of us) that many *differential equation* problems

$$\mathcal{D}(u) = 0$$

can be handled by *variational techniques*, in other words, by considering an *associated real-valued function*

$$\varphi : X \longrightarrow \mathbb{R} ,$$

whose *derivative* is equal to $\mathcal{D}(u)$, and by looking for points of *minimum*, *maximum* or *minimax* (e.g., saddle-like) of φ, so that our given problem reads

$$\varphi'(u) = 0 \quad \text{or} \quad D\varphi(u) = \mathcal{D}(u) = 0 .$$

1 Five Illustrating Problems

Let us start by illustrating our statement through the following five ordinary differential equation problems:

$$\begin{cases} u'' + \frac{1}{2}u = \sin t \ , \quad 0 < t < \pi \\ u(0) = u(\pi) = 0 \end{cases} \tag{P_1}$$

$$\begin{cases} u'' + u = \sin t \ , \quad 0 < t < \pi \\ u(0) = u(\pi) = 0 \end{cases} \tag{P_2}$$

$$\begin{cases} u'' + 2u = \sin t \ , \quad 0 < t < \pi \\ u(0) = u(\pi) = 0 \end{cases} \tag{P_3}$$

$$\begin{cases} u'' - u^3 = 0 \ , \quad 0 < t < \pi \\ u(0) = u(\pi) = 0 \end{cases} \tag{P_4}$$

$$\begin{cases} u'' + u^3 = 0 \ , & 0 < t < \pi \\ u(0) = u(\pi) = 0 \end{cases} \tag{P_5}$$

Except for 5), the following statements about these problems are clear:

1) Problem (P_1) has the solution $u_0(t) = -2\sin t$ (unique);
2) Problem (P_2) has *no* solution;
3) Problem (P_3) has the solution $u_0(t) = \sin t$ (unique);
4) Problem (P_4) has the solution $u_0(t) = 0$ (unique);
5) Problem (P_5) has the solution $u_0(t) = 0$ (not unique).

Indeed, statements $1), 2)$ and 3) are easily checked through elementary methods of solution for second order linear differential equations. Also, the fact that $u_0(t) = 0$ is a solution of both (P_4) and (P_5) is obvious, whereas the fact that (P_4) has no other solution can be seen by multiplying the equation by u and integrating by parts. We then obtain

$$0 = \int_0^\pi u''(t)u(t)\, dt - \int_0^\pi u(t)^4\, dt = -\int_0^\pi (u'(t))^2\, dt - \int_0^\pi u(t)^4\, dt \ ,$$

and it readily follows that $u(t) = 0$. What is not so obvious (unless one uses, for instance, a phase-plane analysis) is that Problem (P_5) has other solutions besides $u(t) = 0$, in fact infinitely many of them.

Now, let us note that each of the Problems $(P_1) - (P_5)$ is of the form

$$\begin{cases} u'' + f(u) = \rho(t) \ , & 0 < t < \pi \\ u(0) = u(\pi) = 0 \end{cases} \tag{P}$$

where $f(s) = \frac{1}{2}s$, s, $2s$, $-s^3$, s^3 (respectively) and $\rho(t) = \sin t$, $\sin t$, $\sin t$, 0, 0 (respectively). And, if $u_0(t)$ is a solution, then by multiplying the equation by an arbitrary smooth function $h(t)$ satisfying $h(0) = h(\pi) = 0$ and integrating by parts, we obtain

$$-\int_0^\pi u_0'(t)h'(t)\, dt + \int_0^\pi f(u_0(t))h(t)\, dt = \int_0^\pi \rho(t)h(t)\, dt \ ,$$

or

$$\int_0^\pi u_0'(t)h'(t)\, dt - \int_0^\pi f(u_0(t))h(t)\, dt + \int_0^\pi \rho(t)h(t)\, dt = 0 \ .$$

It is easy to verify that the above expression says that the directional derivative of φ at u_0 (in the direction of our arbitrary h) equals zero,

$$D\varphi(u_0)\cdot h := \lim_{\delta\to 0} \frac{\varphi(u_0 + \delta h) - \varphi(u_0)}{\delta} = 0 \ ,$$

where φ is the functional given by

$$\varphi(u) := \frac{1}{2}\int_0^\pi (u'(t))^2\, dt - \int_0^\pi F(u(t))\, dt + \int_0^\pi \rho(t)u(t)\, dt\ ,$$

where $F(s) = \int_0^s f(\tau)d\tau$. Therefore, we have shown that a solution of problem (P) is a *critical point* of the functional φ above.

The informal treatment above and the zero boundary condition imposed on (P) suggests the introduction of the Hilbert space

$$X = H_0^1(0,\pi)^1$$

endowed with the inner product and norm given by

$$\langle u,v\rangle := \int_0^\pi u'(t)v'(t)\, dt\ ,\quad ||u|| := \left(\int_0^\pi (u'(t))^2\, dt\right)^{1/2}\ ,$$

and the functional $\varphi : X \longrightarrow \mathbb{R}$ defined by

$$\varphi(u) = \frac{1}{2}||u||^2 - \psi(u)\ ,$$

where

$$\psi(u) := \int_0^\pi F(u(t))\, dt - \int_0^\pi \rho(t)u(t)\, dt\ .$$

At this point, a question that naturally arises is the following:

Question. What *type* of critical point is each solution u_0 presented earlier? What is the *geometry* of the functional φ in each problem $(P_1) - (P_5)$?

Among other topics, the present monograph will answer this and other related questions. It turns out that, in problems (P_1) and (P_4), the solution u_0 is a *global minimum* of φ, whereas u_0 is a *local minimum* of φ in problem (P_5). The (nonobvious) *nonzero* solutions of (P_5) mentioned earlier turn out to be *minimax* points $\widehat{u}$ of φ which can be obtained through the celebrated *mountain-pass theorem* of Ambrosetti and Rabinowitz, and the *symmetric mountain-pass theorem* of Rabinowitz [63]. On the other hand, the solution $u_0(t) = \sin t$ in problem (P_3) turns out to be a *minimax* critical point of φ which could be obtained through the *saddle-point theorem* of Rabinowitz [63]. We provide below a sneak preview of an answer to our question through some pictures (Note: The 3D-picture assigned to Problem (P_5) is somewhat misleading, as the corresponding functional is not bounded from above; keep in

[1] The space $H_0^1(0,\pi)$ consists of all absolutely continuous functions $u : [0,\pi] \to \mathbb{R}$ such that $u(0) = u(\pi) = 0$ and $u' \in L^2(0,\pi)$.

mind that all functionals are defined on an *infinite-dimensional* space). Incidentally, the fact that the linear problem (P_2) has *no* solution (as indicated by the slanted trough in the pictures below) is simply due to the resonant nature of that problem, as the coefficient $\lambda_1 = 1$ of u on the left-hand side of (P_2) is an eigenvalue (the first) of $u'' + \lambda u = 0$, $u(0) = u(\pi) = 0$. As it is well known (from the *Fredholm alternative*), an additional condition is needed for $u'' + \lambda_1 u = \rho(t)$, $u(0) = u(\pi) = 0$ to have a solution, namely, $\int_0^\pi \rho(t) \sin t \, dt = 0$ must hold. The right-hand side $\rho(t) = \sin t$ of (P_2) clearly violates this condition! As we shall see in Chapter 5 (also Chapter 11), there are corresponding conditions for nonlinear problems which were first introduced by Landesman and Lazer in [49].

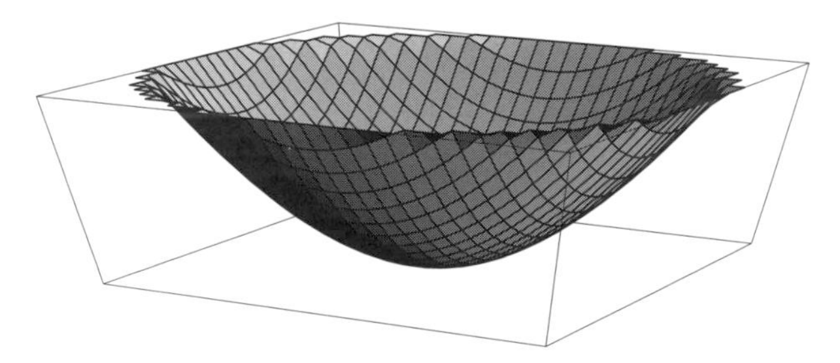

Fig. 1.1. Problems (P_1) and (P_4)

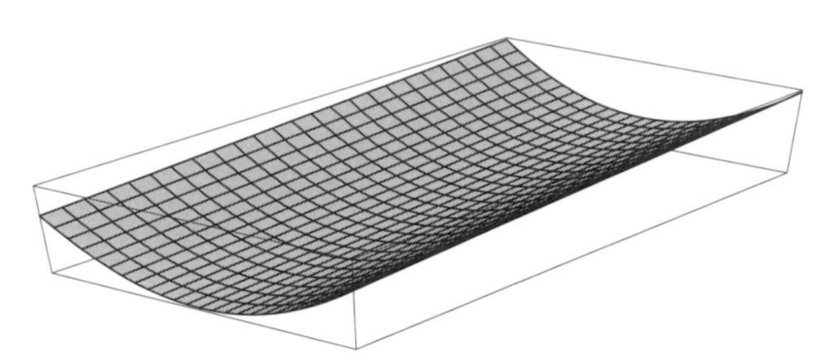

Fig. 1.2. Problem (P_2)

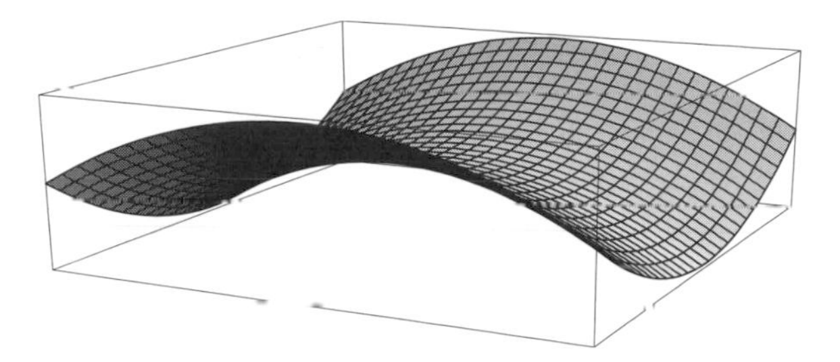

Fig. 1.3. Problem (P_3)

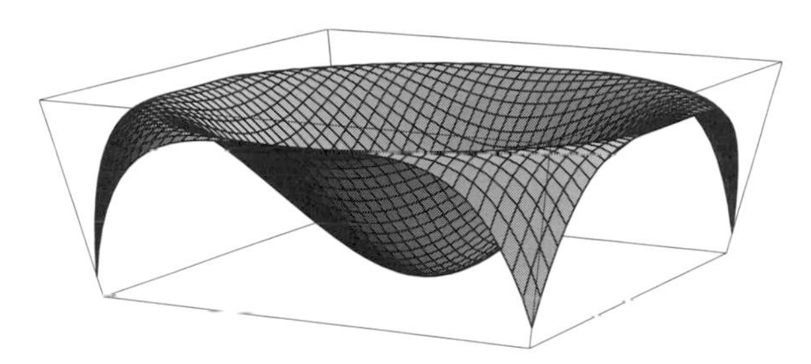

Fig. 1.4. Problem (P_5)

2

Critical Points Via Minimization

1 Basic Results

One of the most basic minimization problems one can pose is the following:

> Given a functional $\varphi : E \longrightarrow \mathbb{R}$ on a Hilbert space E and a closed, convex subset $C \subset E$ on which φ is bounded from below, find $u_0 \in C$ such that
> $$\varphi(u_0) = \inf_{u \in C} \varphi(u) \ .$$

Of course, the problem as stated is much too general and one should be careful and make additional hypotheses! Nowadays, any good calculus student is aware of the fact that a function $\varphi : \mathbb{R} \longrightarrow \mathbb{R}$ which is bounded from below on a closed interval $C \subset \mathbb{R}$ does not necessarily attain its infimum in C. In fact, in learning the classic Weierstrass theorem, the student might have discovered that this delicate question of attaining the infimum is intimately connected to "continuity" of the functional φ and "compactness" of the set C! More precisely, the result that follows is well known and standard in a first course on Topology. For that, we recall that a functional $\varphi : X \longrightarrow \mathbb{R}$ on a topological space X is *lower-semicontinuous (l.s.c.)* if $\varphi^{-1}(a, \infty)$ is open in X for any $a \in \mathbb{R}$ (that is, $\varphi^{-1}(-\infty, a]$ is closed in X for any $a \in \mathbb{R}$). And if X satisfies the first countability axiom (for example, if X is a metric space), then $\varphi : X \longrightarrow \mathbb{R}$ is l.s.c. if and only if $\varphi(\hat{u}) \leq \liminf \varphi(u_n)$ for any $\hat{u} \in X$ and sequence u_n converging to $\hat{u}$.

Theorem 1.1. *Let X be a compact topological space and $\varphi : X \longrightarrow \mathbb{R}$ a lower-semicontinuous functional. Then φ is bounded from below and there exists $u_0 \in X$ such that*

$$\varphi(u_0) = \inf_X \varphi \ .$$

Proof: We can clearly write $X = \cup_{n=1}^{\infty} \varphi^{-1}(-n, \infty)$. Since, by hypothesis, each set $\varphi^{-1}(-n, \infty)$ is open and X is compact, it follows that

$$X = \bigcup_{n=1}^{n_0} \varphi^{-1}(-n, \infty)$$

for some $n_0 \in \mathbb{N}$, hence $\varphi(u) > -n_0$ for all $u \in X$, so that φ is bounded from below. Now, let $c = \inf_X \varphi > -\infty$ and suppose, by contradiction, that $\varphi(u) > c$ for all $u \in X$. Then,

$$X = \bigcup_{n=1}^{\infty} \varphi^{-1}(c + \frac{1}{n}, \infty)$$

and, again, by compactness of X, there exists $k \in \mathbb{N}$ such that $\varphi(u) > c + 1/k$ for all $u \in X$, hence $c \geq c + 1/k$, which is absurd. Therefore, the infimum c must be attained. $\square$

Next, given a Hilbert space E and a functional $\varphi : E \longrightarrow \mathbb{R}$, we recall that φ is *weakly l.s.c.* if it is l.s.c. considering E with its weak topology: in other words, $\varphi(\hat{u}) \leq \liminf \varphi(u_n)$ whenever u_n converges weakly to $\hat{u}$ (written $u_n \rightharpoonup \hat{u}$)[1]. As a consequence of Theorem 1.1, we obtain the following important result which represents a synthesis of the so-called "direct method of the calculus of variations".

Theorem 1.2. *Let E be a Hilbert space (or, more generally, a reflexive Banach space) and suppose that a functional $\varphi : E \longrightarrow \mathbb{R}$ is*

(i) weakly lower-semicontinuous (weakly l.s.c.),
(ii) coercive (that is, $\varphi(u) \longrightarrow +\infty$ as $||u|| \rightarrow \infty$).

Then φ is bounded from below and there exists $u_0 \in E$ such that

$$\varphi(u_0) = \inf_E \varphi \ .$$

Proof: By the coercivity hypothesis (ii), choose $R > 0$ such that $\varphi(u) \geq \varphi(0)$ for all $u \in E$ with $||u|| \geq R$. Since the closed ball $\bar{B}_R$ (of radius R and center at 0) is compact in the weak topology and, by (i), $\varphi : \bar{B}_R \longrightarrow \mathbb{R}$ is l.s.c. in the weak topology, Theorem 1.1 implies the existence of $u_0 \in \bar{B}_R$ such that $\varphi(u_0) = \inf_{\bar{B}_R} \varphi$, hence $\varphi(u_0) = \inf_E \varphi$ by the choice of R. $\square$

[1] $u_n \rightharpoonup \hat{u}$ iff $\langle u_n, h\rangle \longrightarrow \langle \hat{u}, h\rangle$ for all $h \in E$, where $\langle \cdot, \cdot \rangle$ denotes the inner product.

Remark 1.1. In addition to the above conditions, if the given functional $\varphi : E \longrightarrow \mathbb{R}$ is differentiable, then any point u_0 of minimum is a critical point of φ, that is, $\varphi'(u_0) = 0 \in E^*$. This follows from a standard calculus argument (which the reader is invited to recall).

As another consequence of Theorem 1.1, we can now state the following answer to the minimization problem mentioned in the beginning of this section.

Theorem 1.3. *Under the hypotheses* $(i), (ii)$ *of the previous theorem, given any closed, convex subset* $C \subset E$ *there exists* $\hat{u} \in C$ *such that* $\varphi(\hat{u}) = \inf_C \varphi$.

Proof: The proof is essentially the same as that of the previous theorem, if we replace $\bar{B}_R$ [2] by $\bar{B}_R \cap C$ and keep in mind that a closed, convex and bounded subset of E is (again) weakly compact (cf. [36], sections V.3 and V.4). □

Example. Let E be a Hilbert space, $a : E \times E \longrightarrow \mathbb{R}$ a continuous symmetric bilinear form satisfying $a(u, u) \geq \alpha ||u||^2$ for all $u \in E$ (and some $\alpha > 0$) and let $l : E \longrightarrow \mathbb{R}$ be a continuous linear functional. Consider the *quadratic* functional defined by

$$\varphi(u) = \frac{1}{2}a(u, u) - l(u) , \quad u \in E .$$

Then, given any *admissible* set C (that is, a closed, convex subset $C \subset E$), the classic minimization problem

$$\varphi(\hat{u}) = \inf_{u \in C} \varphi(u)$$

has a unique solution. In fact, existence of $\hat{u} \in C$ is guaranteed by Theorem 1.3 by noticing that the quadratic functional φ is coercive and, being continuous and convex, is weakly l.s.c. (see Example B below). Uniqueness follows from the strict convexity of φ in this case. In the special situation in which $a(u, v) = \langle u, v \rangle$ (the inner product of E), we have

$$\varphi(u) = \frac{1}{2}||u||^2 - \langle u, h \rangle , \quad u \in E ,$$

(where $h \in E$ represents the linear functional l, according to the Riesz–Fréchet representation theorem). By *completing the square* in the above expression, that is, by writing $\varphi(u)$ as

[2] Similarly, in this case $R > 0$ is chosen so that $\varphi(u) \geq \varphi(p)$ for all $u \in C$ with $||u|| \geq R$, where $p \in C$ is any fixed point.

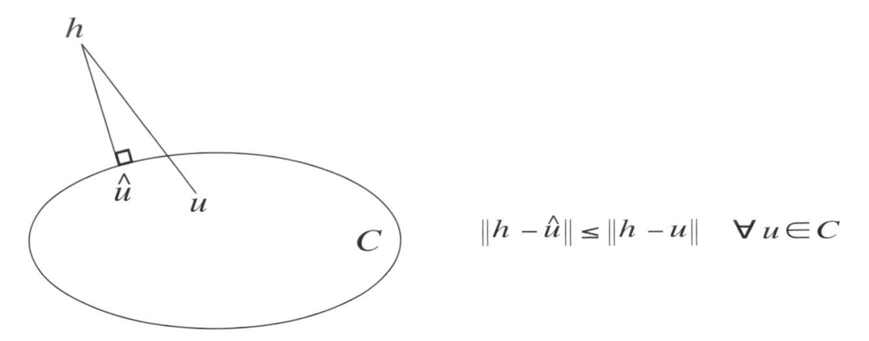

Fig. 2.1.

$$\varphi(u) = \frac{1}{2}||u - h||^2 - \frac{1}{2}||h||^2 \ , \quad u \in E \ ,$$

we see that the point $\hat{u} \in C$ has the geometric characterization of being the projection of h onto the closed, convex set C, $\hat{u} = Proj_C h$. In other words, $\hat{u} \in C$ is such that $||h - \hat{u}|| \leq ||h - u|| \ \ \forall u \in C$.

Example A. If $K : E \longrightarrow E$ is a completely continuous operator on a Hilbert space E, then the functional $\varphi : E \longrightarrow \mathbb{R}$ given below is weakly l.s.c.,

$$\varphi(u) = \langle Ku, u \rangle \ , \quad u \in E \ .$$

In fact, φ is weakly continuous since, in this case, $Ku_n \to Ku$ strongly if $u_n \rightharpoonup u$ weakly, hence $\varphi(u_n) \to \varphi(u)$ if $u_n \rightharpoonup u$ weakly. When K is a positive symmetric operator (continuous but not necessarily completely continuous) the above functional is convex and its lower semicontinuity in the weak topology is then a consequence of its lower semicontinuity, as the next class of examples shows.

Example B. If $\varphi : E \longrightarrow \mathbb{R}$ is a convex, l.s.c. functional on a reflexive Banach space E, then φ is weakly l.s.c.

For the proof it is convenient to introduce the notion of *epigraph* of φ,

$$epi(\varphi) = \{(u, a) \in E \times \mathbb{R} \mid \varphi(u) \leq a\} \ .$$

Then we can use the following easily proven equivalences,

φ is convex (resp. l.s.c., resp. weakly l.s.c.) if and only if
$epi(\varphi)$ is convex (resp. closed, resp. weakly closed),

and recall the fact that a closed, convex subset of a reflexive Banach space ($E \times \mathbb{R}$, in this case) is weakly closed.

Example C. Let $\Omega \subset \mathbb{R}^N$ ($N \geq 1$) be a bounded domain and let $F : \Omega \times \mathbb{R} \longrightarrow \mathbb{R}$ be a function satisfying the so-called *Carathéodory conditions*:

(i) $F(\cdot, s)$ is measurable on Ω for every fixed $s \in \mathbb{R}$,
(ii) $F(x, \cdot)$ is continuous on $\mathbb{R}$ for almost all $x \in \Omega$.

(This implies that the so-called *Nemytskii operator* $u(x) \mapsto F(x, u(x))$ associated with F is well defined on the space of measurable functions $u : \Omega \longrightarrow \mathbb{R}$ (cf. [72], Chapter 2.2)).

Now, we show that under a suitable growth condition, namely

There exist $a, b > 0$ and $1 \leq \alpha < 2N/(N-2)$ if $N \geq 3$ ($1 \leq \alpha < \infty$ if $N = 1, 2$) such that (F_1)

$$|F(x,s)| \leq a|s|^{\alpha} + b\ ,$$

the functional ψ given by the formula

$$\psi(u) = \int_{\Omega} F(x, u(x))\, dx$$

is well defined and weakly continuous on the Sobolev space $H_0^1(\Omega)$.

Indeed, the main point here is that the Sobolev space H_0^1 is compactly embedded in L^p for any $1 \leq p < 2N/(N-2)$, in view of Sobolev embedding Theorem (cf. [1]) and the growth condition (F_1) which implies that the Nemytskii operator F maps the space L^p, for $p \geq \alpha$, into the space $L^{p/\alpha}$ in a continuous manner (cf. the Vainberg theorem in [71]). Therefore, if $u_n \rightharpoonup u$ weakly in H_0^1, then $u_n \to u$ strongly in L^p (for any $1 \leq p < 2N/(N-2)$) and so, by the continuity of the Nemytskii operator just mentioned, it follows that $F(\cdot, u_n) \to F(\cdot, u)$ strongly in $L^{p/\alpha}$, hence strongly in L^1 (since Ω is a bounded domain). In other words, $\psi(u_n) \to \psi(u)$ whenever $u_n \rightharpoonup u$ weakly in H_0^1.

2 Application to a Dirichlet Problem

Let us consider the following nonlinear Dirichlet problem

$$\begin{cases} -\Delta u = f(x,u) & \text{in } \Omega \\ u = 0 & \text{on } \partial\Omega\ , \end{cases} \qquad (P)$$

where $\Omega \subset \mathbb{R}^N$ ($N \geq 1$) is a bounded domain and $f : \Omega \times \mathbb{R} \longrightarrow \mathbb{R}$ is a Carathéodory function (that is, a function satisfying the Carathéodory conditions stated in the previous section).

We shall be interested in finding *weak solutions* of (P), that is, functions $u \in H_0^1(\Omega)$ such that

$$\int_\Omega [\nabla u \cdot \nabla h - f(x,u)h] \, dx = 0 \quad \forall \, h \in H_0^1(\Omega) \, . \tag{2.1}$$

For that, we introduce the growth condition

There exist $c, d > 0$ and $0 \leq \sigma < (N+2)/(N-2)$ if $N \geq 3$ ($0 \leq \sigma < \infty$ if $N = 1, 2$) such that

$$|f(x,s)| \leq c|s|^\sigma + d \tag{f_1}$$

and start by proving the following basic result.

Proposition 2.1. *Let $f : \Omega \times \mathbb{R} \longrightarrow \mathbb{R}$ be a Carathéodory function satisfying condition (f_1). Then, letting $F(x,s) = \int_0^s f(x,\tau) \, d\tau$, the functional*

$$\varphi(u) = \int_\Omega [\frac{1}{2}|\nabla u|^2 - F(x,u)] \, dx \quad , \quad u \in H_0^1(\Omega) \, ,$$

is well defined and, in fact, $\varphi \in C^1(H_0^1, \mathbb{R})$ with

$$\varphi'(u) \cdot h = \int_\Omega [\nabla u \cdot \nabla h - f(x,u)h] \, dx = 0 \quad \forall \, u, h \in H_0^1(\Omega) \, . \tag{2.2}$$

(Therefore, $u \in H_0^1$ is a weak solution of (P) if and only if u is a critical point of φ.)

Proof: We shall always consider the Sobolev space H_0^1 endowed with the norm

$$||u|| = (\int_\Omega |\nabla u|^2 \, dx)^{1/2} \, ,$$

which is equivalent to the usual norm $(||u||_{L^2}^2 + ||\nabla u||_{L^2}^2)^{1/2}$ in view of *Poincaré inequality*,

$$||u||_{L^2}^2 \leq \lambda_1^{-1} ||\nabla u||_{L^2}^2 \quad \forall \, u \in H_0^1 \, ,$$

where $\lambda_1 > 0$ is the first eigenvalue of the problem $-\Delta u = \lambda u$ in Ω, $u = 0$ on $\partial\Omega$ (cf. Exercise 3.5). Therefore, we may write

$$\varphi(u) = \frac{1}{2}||u||^2 - \psi(u) \quad , \quad \psi(u) = \int_\Omega F(x,u) \, dx \, ,$$

and it suffices to show that ψ is well defined and $\psi \in C^1(H_0^1, \mathbb{R})$ with

$$\psi'(u) \cdot h = \int_\Omega f(x,u)h \, dx \quad \forall \, u,h \in H_0^1(\Omega) \tag{2.3}$$

since the functional $q(u) = \frac{1}{2}||u||^2$ is clearly of class C^∞ with $q'(u) \cdot h = \int \nabla u \cdot \nabla h \, dx = \langle u,h \rangle \ \ \forall u,h \in H_0^1$. Now, since f is a Carathéodory function satisfying (f_1), it easily follows that F is also a Carathéodory function and it satisfies the condition (F_1) of the previous section (Example C). Therefore, $\psi : H_0^1 \longrightarrow \mathbb{R}$ is well defined and weakly l.s.c.

In order to show the differentiability of ψ, let $u \in H_0^1$ be fixed and define

$$\delta(h) = \psi(u+h) - \psi(u) - \int_\Omega f(x,u)h \, dx \, .$$

Then

$$\begin{aligned} \delta(h) &= \int_\Omega \int_0^1 \frac{d}{dt} F(x, u+th) \, dt \, dx - \int_\Omega f(x,u)h \, dx \\ &= \int_0^1 \left(\int_\Omega [f(x,u+th) - f(x,u)]h \, dx \right) dt \, , \end{aligned}$$

so that, by Hölder inequality, we obtain the estimate

$$|\delta(h)| \le \int_0^1 ||f(\cdot, u+th) - f(\cdot,u)||_{L^r} ||h||_{L^s} \, dt \, , \tag{2.4}$$

where $r = 2N/(N+2)$ and $s = 2N/(N-2)$. (Here, we only consider the case $N \ge 3$. The cases $N = 1,2$ must be treated separately.) Now, in view of the continuous Sobolev embedding $H_0^1 \hookrightarrow L^s$, we have that $(u+th) \to u$ in L^s as $h \to 0$ in H_0^1. Therefore $f(\cdot, u+th) \to f(\cdot,u)$ in $L^{s/\sigma}$ by Vainberg theorem [71]. And since $r = 2N/(N+2) < s/\sigma$, it follows that

$$f(\cdot, u+th) \longrightarrow f(\cdot,u) \ \text{ in } L^r \, ,$$

hence, from (2.4),

$$\frac{|\delta(h)|}{||h||} \le c \frac{|\delta(h)|}{||h||_{L^s}} \le c \int_0^1 ||f(\cdot,u+th) - f(\cdot,u)||_{L^r} \, dt \longrightarrow 0$$

as $||h|| \to 0$ in view of Lebesgue dominated convergence theorem (also notice that we used the Sobolev inequality $||h||_{L^s} \le c||h||$ in the first inequality above). We have thus shown that $\psi : H_0^1 \longrightarrow \mathbb{R}$ is (Fréchet) differentiable at any $u \in H_0^1$ with derivative $\psi'(u)$ given by (2.3).

Finally, in order to verify the continuity of $\psi' : H_0^1 \longrightarrow (H_0^1)^* = H^{-1}$, we again use the Hölder inequality, Sobolev embedding and the Vainberg theorem as above to obtain

$$\begin{aligned} ||\psi'(u+v) - \psi'(u)||_{H^{-1}} &= \sup_{||h|| \leq 1} |[\psi'(u+v) - \psi'(u)] \cdot h| \\ &\leq c||f(\cdot, u+v) - f(\cdot, u)||_{L^r} \longrightarrow 0 \end{aligned} \tag{2.5}$$

as $v \to 0$ in H_0^1. □

Remark 2.1. Let us denote by $\nabla\varphi : H_0^1 \longrightarrow H_0^1$ the gradient of φ, which is defined via the Riesz–Fréchet representation theorem, that is, $\nabla\varphi(u) \in H_0^1$ is the unique element such that $\varphi'(u) \cdot h = \langle h, \nabla\varphi(u) \rangle$ for all $h \in H_0^1$. Then, it follows from (2.2), (2.3) that $\nabla\varphi(u) = u - T(u)$, where $T : H_0^1 \longrightarrow H_0^1$, $T(u) = \nabla\psi(u)$, is a *compact operator*. Indeed, if $u_n \rightharpoonup u$ weakly in H_0^1, then $u_n \to u$ strongly in L^{s_1} for any s_1 with $1 \leq s_1 < s = 2N/(N-2)$, in view of the *compact* Sobolev embedding $H_0^1 \hookrightarrow L^{s_1}$. Therefore, fixing s_1 so that $\sigma + 1 \leq s_1 < s$ and arguing as in (2.5), we obtain

$$\begin{aligned} ||T(u_n) - T(u)|| &= ||\psi'(u_n) - \psi'(u)||_{H^{-1}} \\ &\leq c||f(\cdot, u_n) - f(\cdot, u)||_{L^{r_1}} \longrightarrow 0 \end{aligned}$$

since $r_1 = s_1/(s_1 - 1) \leq s_1/\sigma$. □

We are now in a position to prove the following existence result for problem (P) :

Theorem 2.2. *Suppose that* $f : \Omega \times \mathbb{R} \longrightarrow \mathbb{R}$ *is a Carathéodory function satisfying conditions* (f_1) *and* (f_2), *where*

There exists $\beta < \lambda_1$ *such that* $\limsup_{|s| \to \infty} \frac{f(x,s)}{s} \leq \beta$ *uniformly for* $x \in \Omega$. $\quad (f_2)$

Then problem (P) *has a weak solution* $u \in H_0^1(\Omega)$.

Proof: In view of Proposition 2.1, we shall find a critical point of the functional $\varphi \in C^1(H_0^1, \mathbb{R})$ given by

$$\varphi(u) = \frac{1}{2}||u||^2 - \psi(u) \quad , \quad \psi(u) = \int_\Omega F(x, u)\, dx \, .$$

As we know, $q(u) = \frac{1}{2}||u||^2$ is weakly l.s.c. and ψ is weakly continuous (cf. Example C of the previous section). Therefore, it follows that

(a) φ **is weakly l.s.c.**

On the other hand, condition (f_2) implies (by l'Hôspital's Rule)

$$\limsup_{|s|\to\infty} \frac{2F(x,s)}{s^2} \le \beta \quad \text{uniformly for } x \in \Omega, \tag{$\hat{f}_2$}$$

and hence, taking β_1 such that $\beta < \beta_1 < \lambda_1$, we can find R_1 such that $F(x,s) \le \frac{1}{2}\beta_1 s^2$ for all $x \in \Omega$, $|s| \ge R_1$. Since (f_1) also gives $F(x,s) \le \gamma_1$ for all $x \in \Omega$, $|s| \le R_1$, we get

$$F(x,s) < \frac{1}{2}\beta_1 s^2 + \gamma_1 \qquad \forall\, x \in \Omega, \forall\, s \in \mathbb{R}\,.$$

This implies the following estimate from below for φ,

$$\varphi(u) \ge \frac{1}{2}\int_\Omega |\nabla u|^2\, dx - \frac{1}{2}\beta_1 \int_\Omega u^2\, dx - \gamma_1|\Omega|\,,$$

which combined with the Poincaré inequality $\lambda_1 \int_\Omega u^2 \le \int_\Omega |\nabla u|^2$ yields

$$\varphi(u) \ge \frac{1}{2}\left(1 - \frac{\beta_1}{\lambda_1}\right)\int_\Omega |\nabla u|^2\, dx - \gamma = \frac{1}{2}a||u||^2 - \gamma\,,$$

where $a = 1 - \frac{\beta_1}{\lambda_1} > 0$ and $\gamma = \gamma_1|\Omega|$. Therefore we also conclude that

(b) φ **is coercive on H_0^1.**

Finally, in view of (a), (b) and Theorem 1.2, it follows that there exists $u_0 \in H_0^1$ such that $\varphi(u_0) = \inf_{H_0^1} \varphi$. Therefore, by Remark 1.1 and Proposition 2.1, u_0 is a critical point of φ. □

Remark 2.2. Theorem 2.2 is essentially due to Hammerstein [46] and Dolph [35] (see also [56], where a weak form of ($\hat{f}_2$) is assumed instead of (f_2)).

3 Exercises

1. Prove Remark 1.1.
2. Provide the details in Example B.
3. Let $F : \mathbb{R} \longrightarrow \mathbb{R}$ be continuous. Show that the functional ψ given by the formula

$$\psi(u) = \int_a^b F(u(x))\, dx$$

is well defined and weakly continuous on the Sobolev space $H_0^1(a,b)$. [Note that no growth condition like (F_1) is imposed on F.]

4. Given $g \in L^2(\Omega)$, with Ω a bounded domain, the functional $\varphi(u) = \frac{1}{2}\int_\Omega |\nabla u|^2\,dx - \int_\Omega g\,u\,dx$, $u \in H_0^1(\Omega)$ provides an illustration of a quadratic functional in the class considered in the Example following Theorem 1.3. As we know, its unique critical point (the minimum of φ) is the weak solution $\hat{u} = h \in H_0^1(\Omega)$ of the linear problem

$$\begin{cases} -\Delta u = g(x) & \text{in } \Omega \\ u = 0 & \text{on } \partial\Omega \,; \end{cases} \qquad (D)$$

which, in turn, is the *representative* via the Riesz–Fréchet theorem of the continuous linear functional

$$u \mapsto \int_\Omega g\,u\,dx\,.$$

Consider the mapping $K : L^2(\Omega) \longrightarrow L^2(\Omega)$ defined by $Kg = \hat{u}$. Show that K is a *compact, self-adjoint* and *positive operator* on $L^2(\Omega)$ with $ker(K) = \{g | Kg = 0\} = \{0\}$.

5. With $K : L^2(\Omega) \longrightarrow L^2(\Omega)$ as in the previous exercise, use the *spectral theorem* to conclude that there exist sequences $(\mu_j)_{j\in\mathbb{N}} \subset (0, +\infty)$ and $(\phi_j)_{j\in\mathbb{N}} \subset L^2(\Omega)$ such that

$$K\phi_j = \mu_j \phi_j\;,$$

where $\mu_1 \geq \mu_2 \geq \cdots \geq \mu_j > 0$, $\lim_{j\to\infty} \mu_j = 0$, and (ϕ_j) is a *complete orthonormal sequence* for $L^2(\Omega)$. Therefore, $L^2(\Omega)$ possesses an *orthonormal basis* consisting of *eigenfunctions* of $-\Delta$ under the Dirichlet boundary condition, $-\Delta\phi_j = \lambda_j\phi_j$ in Ω, $\phi_j = 0$ on $\partial\Omega$, with $\lambda_j = \frac{1}{\mu_j} > 0$. Using the defining relation for the operator K, namely

$$\langle Kg, h\rangle_{H_0^1} = \langle g, h\rangle_{L^2} \quad \forall\, g \in L^2,\ h \in H_0^1,$$

show that $(\phi_j) \subset H_0^1$ is also an *(orthogonal) basis* for $H_0^1(\Omega)$ and obtain the *Poincaré inequality*

$$\lambda_1 \int_\Omega |u|^2\,dx \leq \int_\Omega |\nabla u|^2\,dx \quad \forall\, u \in H_0^1(\Omega)$$

with the best constant $C = \frac{1}{\lambda_1}$.

6. Let $\Omega \subset \mathbb{R}^N$ $(N \geq 1)$ be a bounded, smooth domain and let $f : \overline{\Omega}\times\mathbb{R} \longrightarrow \mathbb{R}$ be a continuous function. Consider the following *Dirichlet* boundary value problem and *Neumann* boundary value problem:

$$\begin{cases} -\Delta u = f(x,u) & \text{in } \Omega \\ u = 0 & \text{on } \partial\Omega \,; \end{cases} \qquad (D)$$

$$\begin{cases} -\Delta u = f(x,u) & \text{in } \Omega \\ \frac{\partial u}{\partial n} = 0 & \text{on } \partial\Omega \,. \end{cases} \tag{N}$$

As we know, a function $u \in H_0^1(\Omega)$ [resp. $u \in H^1(\Omega)$] is a *weak solution* of (D) [resp. (N)] if

$$\int_\Omega \nabla u \cdot \nabla h \, dx = \int_\Omega f(x,u)h \, dx \qquad \forall \, h \in C_0^1(\Omega) \quad [\text{resp. } \forall \, h \in C^1(\overline{\Omega})].$$

Let $u \in C^2(\overline{\Omega})$ be given. Verify that

u is a *classical solution* of (D) [resp. (N)]
$\Leftrightarrow u$ is a *weak solution* of (D) [resp. (N)].

7. Given a continuous function $h : [0,1] \to \mathbb{R}$, consider the boundary value problem

$$\begin{cases} -u'' + u^3 = h(t) & 0 < t < 1 \\ u(0) = u(1) = 0 \,, \end{cases} \tag{3.1}$$

and the functional $\varphi(u) = \int_0^1 [\frac{1}{2}|u'|^2 + \frac{1}{4}u^4 - hu] \, dt$ defined on the Sobolev space $H_0^1(0,1)$.

(*i*) Show that the functional φ is bounded from below, hence (3.1) has a weak solution u_0 minimizing φ;

(*ii*) Show that u_0 is the unique solution of (3.1) [*Hint*: The function $s \mapsto s^3$ is increasing, i.e., $(s_1^3 - s_2^3)(s_1 - s_2) > 0 \; \forall \, s_1 \neq s_2$ in $\mathbb{R}$];

(*iii*) Exhibit a boundary value problem which is more general than (3.1) and where the existence of a unique solution (minimizing the corresponding functional) holds true;

(*iv*) Show that if we replace the nonlinear term u^3 by $-u^3$ in (3.1) then the corresponding functional φ is no longer bounded from below, so that we cannot apply the minimization principle.

8. Prove that the *weak* solution $u_0 \in H_0^1(0,1)$ in the previous exercise is actually a *classical* solution $u_0 \in C_0^2[0,1] := \{u \in C^2[0,1] \mid u(0) = u(1) = 0\}$ of problem (3.1). More generally, given $f : [0,1] \times \mathbb{R} \longrightarrow \mathbb{R}$ continuous, prove that any *weak* solution of the 1-dimensional boundary value problem $-u'' = f(t,u)$, $t \in (0,1)$, $u(0) = u(1) = 0$, is automatically a *classical* solution.

9. Consider the Dirichlet problem

$$\begin{cases} -\Delta u = \lambda \sin \, u + f(x) & \text{in } \Omega \\ u = 0 & \text{on } \partial\Omega \,, \end{cases} \tag{3.2}$$

where $\lambda > 0$, $f \in L^2(\Omega)$ and $\Omega \subset \mathbb{R}^N$ is a bounded domain. Show that, for $\lambda < \lambda_1$, the above problem has a weak solution minimizing the functional

$$\varphi(u) = \int_\Omega \left[\frac{1}{2}|\nabla u|^2 + \lambda(\cos\ u - 1) - fu\right] dx\ , \qquad u \in H_0^1(\Omega)\ .$$

10. Consider the Dirichlet problem

$$\begin{cases} -\mathrm{div}\ (A(x)\nabla u) + c(x)u = f(x) & \text{in } \Omega \\ u = 0 & \text{on } \partial\Omega\ , \end{cases} \tag{3.3}$$

where $\Omega \subset \mathbb{R}^N$ is a bounded domain, $c, f \in L^\infty(\Omega)$, and $A(x)$ is an $N \times N$ real *symmetric* matrix with components in $L^\infty(\Omega)$ which is *uniformly positive definite*, i.e., $A(x)\xi \cdot \xi \geq \delta|\xi|^2$ for all $x \in \Omega$, $\xi \in \mathbb{R}^N$ (and some $\delta > 0$).

(i) Show that the problem (3.3) is *variational* in the sense that its weak solutions are the critical points of a suitable functional $\varphi : H_0^1(\Omega) \longrightarrow \mathbb{R}$.

(ii) Assume $\delta > \frac{\|c\|_{L^\infty}}{\lambda_1}$. Show that $\widehat{u} \in H_0^1(\Omega)$ is a weak solution of (3.3) if and only if $\widehat{u}$ is a *global minimum* of φ.

3

The Deformation Theorem

1 Preliminaries

Let $\varphi : X \longrightarrow \mathbb{R}$ be a C^1 functional on a Banach space X. A number $c \in \mathbb{R}$ is called a *critical value* of φ if $\varphi(u) = c$ for some critical point $u \in X$. The set of all *critical points at the level* c is denoted by K_c:

$$K_c = \{u \in X \mid \varphi'(u) = 0,\ \varphi(u) = c\} .$$

Also, we shall denote by φ^c the set of all points in X at levels $\leq c$, that is,

$$\varphi^c = \{u \in X \mid \varphi(u) \leq c\} .$$

A basic ingredient in the topological methods is the so-called *deformation theorem*. Roughly, it says when (and how) we can deform φ^{c_1} into φ^{c_2}, for $c_1 > c_2$ (or $c_1 < c_2$). Since X is not a Hilbert space in general and since we are only assuming φ to be of class C^1, we shall need to use the notion of a *pseudo-gradient*, due to Palais [60].

Definition 1. A *pseudo-gradient field* for $\varphi \in C^1(X, \mathbb{R})$ is a locally Lipschitzian mapping $v : Y \longrightarrow X$, where $Y = \{u \in X | \varphi'(u) \neq 0\}$, satisfying the following conditions:

$$||v(u)|| \leq 2||\varphi'(u)|| \tag{1.1}$$

$$\varphi'(u) \cdot v(u) \geq ||\varphi'(u)||^2 \tag{1.2}$$

for all $u \in Y$. In what follows we shall use the fact that any functional $\varphi \in C^1(X, \mathbb{R})$ has a pseudo-gradient (see a proof in [75]).

If X is a Hilbert space and $\varphi \in C^1(X,\mathbb{R})$ has a locally Lipschitzian derivative $\varphi' : X \longrightarrow X^*$, then the *gradient* of φ (when restricted to Y), $\nabla\varphi : Y \longrightarrow X$, is clearly a pseudo-gradient for φ. (Recall that $\nabla\varphi : X \longrightarrow X$ is defined through the Riesz–Fréchet representation theorem: $\nabla\varphi(u) \in X$ is the unique element verifying $\varphi'(u) \cdot h = \langle h, \nabla\varphi(u)\rangle$ for all $u \in X$.)

2 Some Versions of the Deformation Theorem

There are versions of the deformation theorem due to Palais [60, 59] and Clark [25], among others. In all those versions the functional φ is assumed to satisfy some compactness condition. Here, we shall be interested in Clark's versions, which will be stated for the general case of a Banach space X. Our presentation follows [74] (cf. also [75]), where the starting point is a *quantitative version of the deformation theorem* without a Palais–Smale type condition on φ, a result due to Willem [74].

Theorem 2.1. *Let $\varphi : X \longrightarrow \mathbb{R}$ be a C^1 functional on a Banach space X. Let $S \subset X$, $c \in \mathbb{R}$ and $c, \delta > 0$ be such that*

$$||\varphi'(u)|| \geq \frac{4\epsilon}{\delta} \tag{2.1}$$

for all $u \in \varphi^{-1}[c-2\epsilon, c+2\epsilon] \cap S_{2\delta}$.[1] *Then, there exists a continuous mapping $\eta \in C([0,1] \times X, X)$ such that, for any $u \in X$ and $t \in [0,1]$ one has:*

(i) $\eta(0,u) = u$,
(ii) $\eta(t,u) = u$ *if* $u \notin \varphi^{-1}[c-2\epsilon, c+2\epsilon] \cap S_{2\delta}$,
(iii) $\eta(1, \varphi^{c+\epsilon} \cap S) \subset \varphi^{c-\epsilon} \cap S_\delta$,
(iv) $\eta(t,\cdot) : X \longrightarrow X$ *is a homeomorphism.*

Proof: Let $A = \varphi^{-1}[c-2\epsilon, c+2\epsilon] \cap S_{2\delta}$, $B = \varphi^{-1}[c-\epsilon, c+\epsilon] \cap S_\delta$ and $Y = \{u \in X \mid \varphi'(u) \neq 0\}$, so that $B \subset A \subset Y$. Also, let $v : Y \longrightarrow X$ be a pseudo-gradient for φ and consider a locally Lipschitzian mapping $\rho : X \longrightarrow \mathbb{R}$ such that $0 \leq \rho \leq 1$ and

$$\begin{cases} \rho = 1 \ \ in\ B \\ \rho = 0 \ \ in\ X\backslash A\,. \end{cases}$$

Then, define the following locally Lipschitzian mapping $f : X \longrightarrow X$,

$$f(u) = -\rho(u)\frac{v(u)}{||v(u)||}\,.$$

[1] Given a subset $S \subset X$ and $\alpha > 0$, we let S_α denote the closed α-neighborhood of S defined by $S_\alpha = \{u \in X \mid \mathrm{dist}(u,S) \leq \alpha\}$.

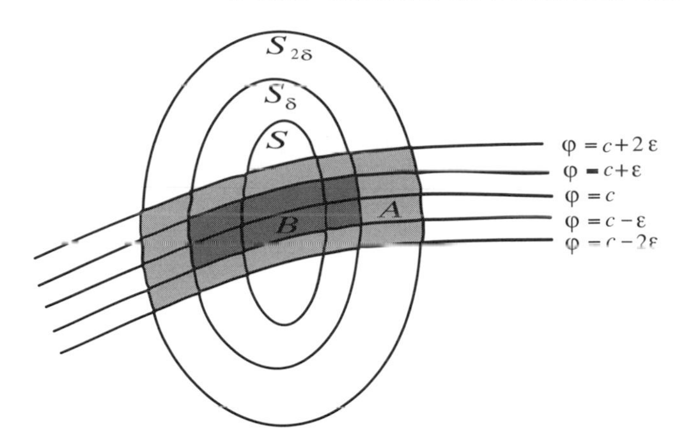

Fig. 3.1.

Since $||f(u)|| \leq 1$ for all $u \in X$, it follows that the Cauchy problem

$$\begin{cases} \dfrac{dw}{dt} &= f(w) \\ w(0) &= u\,. \end{cases}$$

has (for each given $u \in X$) a unique solution defined for all $t \geq 0$. Let $\eta : [0,1] \times X \longrightarrow X$ be defined by

$$\eta(t,u) = w(\delta t, u)\ .$$

Then, it is easy to see that (i), (ii) and (iv) are satisfied. In order to verify (iii) we note that, for $t \geq 0$,

$$||w(t,u) - u|| \leq \int_0^t ||f(w(\tau,u)|| \, d\tau \leq t\ ,$$

so that $w(t,S) \subset S_\delta$ for all $t \in [0,\delta]$, that is,

$$\eta(t,S) \subset S_\delta \quad \text{for all } t \in [0,1]\ . \tag{2.2}$$

Also note that, for each fixed $u \in X$, the function $t \mapsto \varphi(w(t,u))$ is decreasing since

$$\begin{aligned} \frac{d}{dt}\varphi(w(t,u)) &= \varphi'(w(t,u)) \cdot \frac{dw}{dt} \\ &= \varphi'(w(t,u)) \cdot f(w(t,u)) \\ &= -\rho(w(t,u))\varphi'(w(t,u)) \cdot \frac{v(w(t,u))}{||v(w(t,u))||} \\ &\leq -\rho(w(t,u))\frac{||\varphi'(w(t,u))||^2}{||v(w(t,u))||} \\ &\leq 0\ , \end{aligned} \tag{2.3}$$

where we have used (1.2) in the first inequality above. Now, let $u \in \varphi^{c+\epsilon} \cap S$ be given. We have two cases to consider.

(*a*) If $\varphi(w(\hat{t}, u)) < c - \epsilon$ for some $\hat{t} \in [0, \delta)$, then $\varphi(\eta(1, u)) = \varphi(w(\delta, u)) \leq \varphi(w(\hat{t}, u)) < c - \epsilon$, hence $\eta(1, u) \in \varphi^{c-\epsilon} \cap S_\delta$ in view of (2.2).

(*b*) In the other case we have $w(t, u) \in \varphi^{-1}[c - \epsilon, c + \epsilon] \cap S_\delta = B \ \forall t \in [0, \delta]$, and so, using (2.3), (1.1) and the fact that $\rho = 1$ on B, we obtain

$$\varphi(w(\delta, u)) = \varphi(u) + \int_0^\delta \frac{d}{dt}\varphi(w(t, u))\, dt \leq \varphi(u) - \int_0^\delta \frac{1}{2}||\varphi'(w(t, u))||\, dt$$

$$\leq c + \epsilon - \frac{1}{2}\frac{4\epsilon}{\delta}\delta = c - \epsilon \ ,$$

where the assumption (2.1) was used in the last inequality. Therefore, in either case *(a)* or *(b)*, we have shown that $\eta(1, u) = w(\delta, u) \in \varphi^{c-\epsilon} \cap S_\delta$ if $u \in \varphi^{c+\epsilon} \cap S$. The proof is complete □

As a consequence of Theorem 2.1 we obtain the following first version of the deformation theorem, due to Clark [25].

Theorem 2.2. *Let $\varphi \in C^1(X, \mathbb{R})$ satisfy the* Palais–Smale condition*:*

Any sequence (u_n) such that $\varphi(u_n)$ is bounded and $\varphi'(u_n) \to 0$ possesses a convergent subsequence. (PS)

If $c \in \mathbb{R}$ is not *a critical value of φ, then, for every $\epsilon > 0$ sufficiently small, there exists $\eta \in C([0, 1] \times X, X)$ such that, for any $u \in X$ and $t \in [0, 1]$ one has:*

(i) $\eta(0, u) = u$,
(ii) $\eta(t, u) = u$ *if* $u \notin \varphi^{-1}[c - 2\epsilon, c + 2\epsilon]$,
(iii) $\eta(1, \varphi^{c+\epsilon}) \subset \varphi^{c-\epsilon}$,
(iv) $\eta(t, \cdot) : X \longrightarrow X$ *is a homeomorphism.*

Proof: There must exist constants $\alpha, \beta > 0$ such that $||\varphi'(u)|| \geq \beta$ whenever $u \in \varphi^{-1}[c - 2\alpha, c + 2\alpha]$ since, otherwise, one would have a sequence (u_n) satisfying

$$c - \frac{1}{n} \leq \varphi(u_n) \leq c + \frac{1}{n} \ , \quad ||\varphi'(u_n)|| \leq \frac{1}{n} \ ,$$

so that, in view of the condition (PS), the value c would be a critical value of φ, contradicting our assumption.

Now, the result follows from Theorem 2.1 with $S = X$, $\epsilon \in (0, \alpha]$ given and $\delta = 4\epsilon/\beta$. □

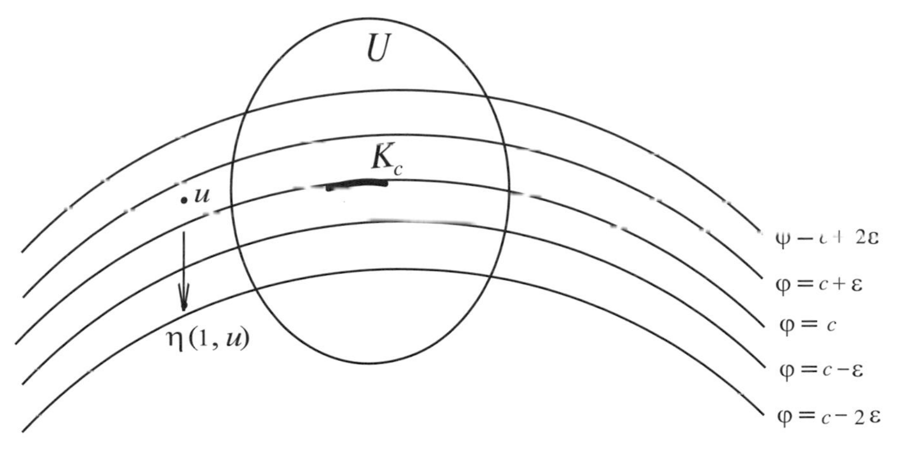

Fig. 3.2.

Remark 2.1. It is clear from the above proof that Theorem 2.2 holds true under the following *weaker* compactness condition introduced by Brézis–Coron–Nirenberg in [18]:

> If a sequence (u_n) is such that $\varphi(u_n) \to c$ and $\varphi'(u_n) \to 0$ then c is a critical value of φ. $(BCN)_c$

This condition may be useful in certain situations where the functional φ is not coercive, as we shall see in the next section. For now, as a simple example, we observe that a function $\varphi : \mathbb{R} \longrightarrow \mathbb{R}$ which is periodic does not satisfy (PS), but it satisfies $(BCN)_c$ for every $c \in \mathbb{R}$.

Next, let us consider a second (more general) version of the deformation theorem, also due to Clark [25].

Theorem 2.3. *Let $\varphi \in C^1(X, \mathbb{R})$ satisfy the Palais–Smale condition (PS). Given $c \in \mathbb{R}$ and an open neighborhood U of K_c, then, for any $\epsilon > 0$ sufficiently small, there exists $\eta \in C([0,1] \times X, X)$ such that (for any $u \in X$ and $t \in [0,1]$):*

(i) $\eta(0, u) = u$,
(ii) $\eta(t, u) = u$ *if* $u \notin \varphi^{-1}[c - 2\epsilon, c + 2\epsilon]$,
(iii) $\eta(1, \varphi^{c+\epsilon} \backslash U) \subset \varphi^{c-\epsilon}$,
(iv) $\eta(t, \cdot) : X \longrightarrow X$ *is a homeomorphism.*

Proof: Let $S = X \backslash U$. Then there must exist $\epsilon, \delta > 0$ such that $||\varphi'(u)|| \geq 4\epsilon/\delta$ wherever $u \in \varphi^{-1}[c - 2\epsilon, c + 2\epsilon] \cap S_{2\delta}$ since, otherwise, one would have a sequence (u_n) satisfying

$$u_n \in S_{\frac{2}{\sqrt{n}}} \quad , \quad c - \frac{2}{n} \leq \varphi(u_n) \leq c + \frac{2}{n} \quad , \quad ||\varphi'(u_n)|| \leq \frac{4}{\sqrt{n}} \; .$$

Then, in view of the condition (PS), one would have a convergent subsequence $u_{n_k} \to u$, with $u \in S \cap K_c$. But this is a contradiction, since $S = X \backslash U$ and $K_c \subset U$. Therefore, the result follows from Theorem 2.1. □

Remark 2.2. We should note that the Palais–Smale condition (PS) implies that the (possibly empty) critical set $K_c = \{u \in X \mid \varphi'(u) = 0 \; , \; \varphi(u) = c\}$ is a compact set. In the case $K_c = \emptyset$ (i.e., c is not a critical value), we can take $U = \emptyset$ in Theorem 2.3 and obtain Theorem 2.2.

3 A Minimum Principle and an Application

As a first application of the theorem of deformation, let us now obtain an important minimum principle, which can be useful in situations where the functional φ is not coercive (compare with Theorem 2.1.2).

Theorem 3.1. *Let $\varphi \in C^1(X, \mathbb{R})$ where X is a Banach space. Assume that*

(i) φ is bounded from below, $c = \inf_X \varphi$,
(ii) φ satisfies $(BCN)_c$.

Then there exists $u_0 \in X$ such that $\varphi(u_0) = c = \inf_X \varphi$ (hence, c is a critical value of φ).

Proof: Let us assume, by negation, that c is not a critical value of φ. Then, Theorem 2.2 implies the existence of $\epsilon > 0$ and $\eta \in C([0,1] \times X, X)$ satisfying $\eta(1, \varphi^{c+\epsilon}) \subset \varphi^{c-\epsilon}$. This is a contradiction since $\varphi^{c-\epsilon} = \emptyset$ (as $c = \inf_X \varphi$). □

Next, we consider an application to the following nonlinear Neumann problem

$$\begin{cases} -\Delta u \; = \; f(u) + \rho(x) \quad \text{in } \Omega \\ \dfrac{\partial u}{\partial n} = 0 \quad \text{on } \partial\Omega \; , \end{cases} \tag{3.1}$$

where $\Omega \subset \mathbb{R}^N$ $(N \geq 1)$ is a bounded smooth domain, $f : \mathbb{R} \longrightarrow \mathbb{R}$, a continuous p-periodic function, and $\rho \in L^2(\Omega, \mathbb{R})$ satisfy the conditions [2]

$$\int_0^p f(s)\, ds = 0 \; , \quad \int_\Omega \rho(x)\, dx = 0 \; . \tag{3.2}$$

[2] As is well known, the second condition is necessary in the linear case $f = 0$.

This situation is the PDE analog of that considered in [74, Section II] for the forced pendulum.

We are interested in finding *weak solutions* of (3.1), in other words, functions $u \in H^1(\Omega)$ satisfying

$$\int_\Omega [\nabla u \cdot \nabla h - f(u)h - \rho h]\, dx = 0 \tag{3.3}$$

for all $h \in H^1(\Omega)$. Here, the Sobolev space $H^1(\Omega)$ is equipped with its usual inner product

$$\langle u, v\rangle = \int_\Omega [uv + \nabla u \cdot \nabla v]\, dx \ ,$$

and we will consider the functional φ given by the formula

$$\varphi(u) = \int_\Omega [\frac{1}{2}|\nabla u|^2 - F(u) - \rho u]\, dx \ , \tag{3.4}$$

where, again, $F(s) = \int_0^s f(\tau)d\tau$.

Proposition 3.2. *The functional $\varphi : H^1(\Omega) \longrightarrow \mathbb{R}$ given above is well defined. Moreover, φ is bounded from below and is of class C^1 with*

$$\varphi'(u) \cdot h = \int_\Omega [\nabla u \cdot \nabla h - f(u)h - \rho h]\, dx \tag{3.5}$$

for all $u, h \in H^1(\Omega)$ (Therefore, $u \in H^1$ is a weak solution *of* (3.1) *if and only if u is a* critical point *of φ).*

Proof: Since the continuous function $f : \mathbb{R} \longrightarrow \mathbb{R}$ is p-periodic and satisfies (3.2), it follows that $F : \mathbb{R} \longrightarrow \mathbb{R}$ is also p-periodic, so that $|F(s)| \leq A$ for some $A > 0$ and φ is well defined on $H^1(\Omega)$ (More generally, and in a similar manner to Proposition 2.2.1, the functional φ will be well defined on $H^1(\Omega)$ provided that $f : \Omega \times \mathbb{R} \longrightarrow \mathbb{R}$ is a Carathéodory function satisfying the growth condition (f_1).)

Now, in order to show that φ is bounded from below, we decompose $X = H^1(\Omega)$ as

$$X = X_0 \oplus X_1 \ ,$$

where $X_1 = \mathbb{R} = span\{1\}$ is the subspace of constant functions and $X_0 = (\text{span}\,\{1\})^\perp = \{\ v \in X \mid \int_\Omega v\ dx = 0\ \}$ is the space of functions in $H^1(\Omega)$ with *mean-value zero.* Also, we observe that the following *Poincaré inequality* holds for functions in X_0, where c is a positive constant (cf. Exercise 5 below):

$$\int_\Omega v^2\, dx \leq c \int_\Omega |\nabla v|^2\, dx \quad \forall\ v \in X_0\ . \tag{3.6}$$

Therefore, writing $u = v + w$ in (3.4), with $v \in X_0$, $w \in X_1$, we obtain

$$\varphi(u) = \varphi(v+w) = \int_\Omega [\frac{1}{2}|\nabla v|^2 - F(v+w) - \rho v]\, dx$$
$$\geq \frac{1}{2}||\nabla v||^2_{L^2} - A|\Omega| - ||\rho||_{L^2}||v||_{L^2} \ ,$$

so that, in view of (3.6),

$$\varphi(u) \geq \frac{1}{2}||\nabla v||^2_{L^2} - C - D||\nabla v||_{L^2} \tag{3.7}$$

for some constants $C, D > 0$, which shows that φ is bounded from below.

Finally, we omit the proof that $\varphi \in C^1(H^1(\Omega), \mathbb{R})$ with the derivative given by (3.5) since it is analogous to that of Proposition 2.2.1. □

Remark 3.1. Again, as in Remark 2.2.1, the gradient mapping $\nabla\varphi : H^1 \longrightarrow H^1$ is of the form $\nabla\varphi(u) = u - T(u)$, where $T : H^1 \longrightarrow H^1$ is a compact operator.

Theorem 3.3. *Assume* (3.2) *with* $f \in C(\mathbb{R}, \mathbb{R})$ *a* p*-periodic function and* $\rho \in L^2(\Omega, \mathbb{R})$*. Then Problem* (3.1) *has a weak solution* $u \in H^1(\Omega)$*.*

Proof: In view of the previous proposition, we shall find a critical point of the functional $\varphi \in C^1(H^1, \mathbb{R})$ given in (3.4). To start, since (3.2) implies that F is p-periodic and $\int_\Omega \rho dx = 0$, we note that the functional φ is also p-periodic, that is,

$$\varphi(u + p) = \varphi(u) \quad \forall\ u \in H^1 \ .$$

Next, we claim that

$$\varphi \ \text{ satisfies } (BCN)_c \text{ for all } c \in \mathbb{R} \ . \tag{$*$}$$

Indeed, let (u_n) be such that $\varphi(u_n) \to c$ and $\varphi'(u_n) \to 0$. Write $u_n = v_n + w_n$ with $v_n \in X_0$ and $w_n \in X_1 = \mathbb{R}$. From (3.7) and the boundedness of $\varphi(u_n)$, it follows that $||\nabla v_n||_{L^2}$ is bounded, hence (v_n) is bounded in H^1 in view of the Poincaré inequality (3.6):

$$||v_n|| \leq C \ . \tag{3.8}$$

On the other hand, letting $\widehat{w}_n \in [0, p)$ be such that $w_n = \widehat{w}_n (mod\ p)$ and defining $\widehat{u}_n = v_n + \widehat{w}_n$, we obtain that

$$\varphi(\widehat{u}_n) = \varphi(u_n) \to c \ , \quad \varphi'(\widehat{u}_n) = \varphi'(u_n) \to 0 \ .$$

Since $||\widehat{u}_n||$ is bounded, in view of (3.8) and the definition of $\widehat{w}_n$, we may assume (passing to a subsequence, if necessary) that $\widehat{u}_n \rightharpoonup \widehat{u}$ weakly in H^1, for some $\widehat{u} \in H^1$. And, since $\widehat{u}_n = \nabla\varphi(\widehat{u}_n) + T(\widehat{u}_n)$, with T a compact operator (cf. Remark 3.1), we conclude that

$$\widehat{u}_n \to 0 + T(\widehat{u})$$

strongly in H^1, that is,

$$\widehat{u}_n \to \widehat{u}$$

strongly in H^1. Therefore, it follows that $\varphi(\widehat{u}) = c$ and $\varphi'(\widehat{u}) = 0$, so that c is a critical value of φ and our claim $(*)$ is true.

Finally, in view of $(*)$ and the fact that φ is bounded from below (cf. (3.7)), we can use Theorem 3.1 to conclude that there exists $u_0 \in H^1$ such that $\varphi(u_0) = \inf_{H^1} \varphi$, hence u_0 is a critical point of φ by Remark 2.1.1. □

4 Exercises

1. Imitate the proof of Theorem 2.1 to show the following simple version of the deformation theorem:

Let E be a Hilbert space and $\varphi : E \longrightarrow \mathbb{R}$ a functional of class C^1 with $\nabla\varphi \in Lip_{loc}(E)$. If $\nabla\varphi(u) \neq 0 \quad \forall u \in E$, then, for any $\epsilon > 0$ sufficiently small, there exists $\eta \subset C([0,1] \times X, X)$ satisfying properties $(i) - (iv)$ of Theorem 2.2.

2. Verify that $(PS)_c \Rightarrow (Ce)_c$ and $(BCN)_c$, where $(Ce)_c$ is the following compactness condition due to Cerami [24]:

If (u_n) is such that $\varphi(u_n) \to c$ and $(1 + \|u_n\|)\|\varphi'(u_n)\| \to 0$, then (u_n) has a convergent subsequence. $(Ce)_c$

3. If $\dim(X) = 1$, decide whether $(PS)_c$ and $(Ce)_c$ are equivalent for a given $\varphi : \mathbb{R} \longrightarrow \mathbb{R}$ of class C^1 (clearly, neither condition is equivalent to $(BCN)_c$).

4. Same question if $\dim(X) = 2$, i.e., for given C^1 functions $\varphi : \mathbb{R}^2 \longrightarrow \mathbb{R}$.

5. Prove the Poincaré inequality (3.6), that is, show that there exists $c > 0$ such that

$$\int_\Omega |u - \overline{u}|^2 \, dx \leq c \int_\Omega |\nabla u|^2 \, dx \qquad \forall \, u \in H^1(\Omega) \; ,$$

where $\overline{u} = \frac{1}{|\Omega|} \int_\Omega u \, dx \in \mathbb{R}$ is the average of u over Ω. Can you guess what is the best (smallest) constant $c > 0$?

4

The Mountain-Pass Theorem

1 Critical Points of Minimax Type

Roughly speaking, the basic idea behind the so-called *minimax method* is the following:

> Find a critical value of a functional $\varphi \in C^1(X,\mathbb{R})$ as a *minimax* (or *maximin*) value $c \in \mathbb{R}$ of φ over a suitable class $\mathcal{A}$ of subsets of X:
>
> $$c = \inf_{A \in \mathcal{A}} \sup_{u \in A} \varphi(u) \ .$$

Example A. Perhaps one of the first examples using a minimax technique is due to E. Fischer (1905) through a well-known minimax characterization of the eigenvalues of a real, symmetric $n \times n$ matrix M (cf. [33], pp. 31 and 47):

$$\lambda_k = \inf_{\{X_{k-1}\}} \sup_{x \perp X_{k-1}, |x|=1} (Mx|x) \ ,$$

$$\lambda_{-k} = \sup_{\{X_{k-1}\}} \inf_{x \perp X_{k-1}, |x|=1} (Mx|x) \ .$$

Here, the eigenvalues are numbered so that $\lambda_{-1} \leq \cdots \leq \lambda_{-k} \leq \cdots \leq 0 \leq \cdots \leq \lambda_k \leq \cdots \leq \lambda_1$. Also we are denoting by $(\cdot|\cdot)$ (resp. $|\cdot|$) the usual inner product (resp. norm) in $X = \mathbb{R}^n$, and by $X_j \subset X$ an arbitrary subspace of dimension j. It should be noted that a characterization which is *dual* to the above characterization also holds true, namely:

$$\lambda_k = \sup_{\{X_k\}} \inf_{x \in X_k, |x|=1} (Mx|x) \ ,$$

$$\lambda_{-k} = \inf_{\{X_k\}} \sup_{x \in X_k, |x|=1} (Mx|x) \ .$$

Example B. A similar characterization can be obtained for the eigenvalues of a compact, symmetric operator $T : X \longrightarrow X$ on a Hilbert space X. This is part of the so-called *Hilbert–Schmidt theory.*

Example C. A *topological* analogue of such minimax schemes was developed by L. Lusternik and L. Schnirelman from 1925 to 1947. This is known as the (classical) *Lusternik–Schnirelman theory.* It was originally based on the topological notion of *category* Cat (A, X) of a closed subset A of a metric space X. By definition, Cat (A, X) is the smallest number of closed, contractible subsets of X which is needed to cover A (see [53, 54]).

In this context, given a functional $\varphi \in C^1(X, \mathbb{R})$ over, say, a differentiable Riemannian manifold X, the idea is to show that the following values are critical values of φ:

$$c_k = \inf_{A \in \mathcal{A}_k} \sup_{x \in A} \varphi(x) \ , \quad k = 1, 2, \dots \ ,$$

where $\mathcal{A}_k := \{ A \subset X \mid A \text{ is closed, Cat}\,(A, X) \geq k \}$. For example, since Cat $(S^n, S^n) = 2$, one obtains, for a given functional $\varphi \in C^1(S^n, \mathbb{R})$, that

$$c_1 \leq c_2 = c_3 = \cdots \ ,$$

and, in this case, $c_1 = \inf \varphi$, $c_2 = \sup \varphi$. Of course this gives us no new information in this case since we know that $\inf \varphi$ and $\sup \varphi$ are attained on the compact manifold S^n and, therefore, are critical values of φ. However, if $\varphi \in C^1(S^n, \mathbb{R})$ is an *even* functional, one obtains more critical values, as shown by the following classical theorem due to Lusternik (1930):

Theorem 1.1. *([53]) Let $\varphi \in C^1(S^n, \mathbb{R})$ be given. If φ is* even, *then it has at least $(n + 1)$ distinct pairs*[1] *of critical points.*

The main idea here is that an *even* functional on S^n can be considered as a functional on the real projective space $\mathbb{RP}^n$ (obtained by identification of the antipodal points in S^n), and the topology of $\mathbb{RP}^n$ is much richer than that of S^n. In fact, it can be shown that Cat $(\mathbb{RP}^n, \mathbb{RP}^n) = n + 1$ (cf. [68]) so that, in this case, one obtains $(n + 1)$ critical values (possibly repeated):

$$c_1 \leq c_2 \leq \cdots \leq c_{n+1}.$$[2]

Another way to interpret Lusternik's multiplicity result is to consider it as a consequence of the *symmetry* of the problem (*evenness* of φ, in this

[1] Clearly, since φ is even, its critical points occur in pairs.

[2] Moreover, if $c_j = c_{j+k}$ for some $j, k \geq 1$, it can be shown that the category of the critical set K_c is at least $k + 1$.

case). This question of *multiplicity versus symmetry* will be tackled in a future chapter.

2 The Mountain-Pass Theorem

As already mentioned in the beginning of this chapter, the basic idea behind the minimax method is to *minimaximize* (or *maximinimize*) a given functional φ over a *suitable class* of subsets of X. In particular, such a *suitable class* can be chosen to be *invariant* under the deformation $\eta(t, \cdot)$ given in the deformation theorem 3.2.3.

In this section we will present a first illustration of the minimax method which has proven to be a powerful tool in the attack of many problems on differential equations. It is the celebrated *mountain-pass theorem* of Ambrosetti and Rabinowitz [9]:

Theorem 2.1. *Let X be a Banach space and $\varphi \in C^1(X, \mathbb{R})$ be a functional satisfying the Palais–Smale condition (PS) (or, more weakly, $(BCN)_c$).*[3] *If $e \in X$ and $0 < r < ||e||$ are such that*

$$a =: \max\{\varphi(0), \varphi(e)\} < \inf_{||u||=r} \varphi(u) =: b \ , \tag{2.1}$$

then

$$c = \inf_{\gamma \in \Gamma} \sup_{t \in [0,1]} \varphi(\gamma(t))$$

is a critical value of φ with $c \geq b$. (Here, Γ is the set of paths joining the points 0 and e, that is, $\Gamma = \{\gamma \in C([0,1], X) \mid \gamma(0) = 0, \ \gamma(1) = e\}$.)

Proof: First note that $\gamma([0,1]) \cap \partial B_r$ is *nonempty* for any given $\gamma \in \Gamma$, since $\gamma(0) = 0$, $\gamma(1) = e$ and $0 < r < ||e||$ by assumption. Therefore,

$$\max_{t \in [0,1]} \varphi(\gamma(t)) \geq b = \inf_{\partial B_r} \varphi \ ,$$

so that $c \geq b$.

Let us assume, by negation, that c is not a critical value. Then, by the deformation theorem 3.2.2, there exist $0 < \epsilon < \frac{b-a}{2}$ (recall that $a < b$ by (2.1) and $\eta \in C([0,1] \times X, X))$ such that

$$\eta(t, u) = u \ \ if \ \ u \notin \varphi^{-1}([c - 2\epsilon, c + 2\epsilon]) \ , \ \ t \in [0,1] \ , \tag{2.2}$$

[3] Recall Remark 3.2.1. One could also use $(Ce)_c$ (cf. [67]).

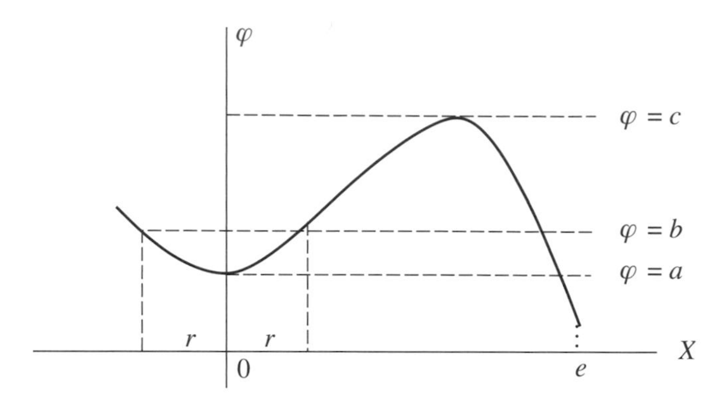

Fig. 4.1.

$$\eta(1, \varphi^{c+\epsilon}) \subset \varphi^{c-\epsilon} . \tag{2.3}$$

Now, by definition of c as an infimum over Γ, we can choose $\gamma \in \Gamma$ such that

$$\max_{t\in[0,1]} \varphi(\gamma(t)) \leq c + \epsilon \tag{2.4}$$

and define the path $\widehat{\gamma}(t) = \eta(1, \gamma(t))$. In view of (2.2) and the fact that $2\epsilon < b - a$, it follows that $\widehat{\gamma} \in \Gamma$ (indeed, $\widehat{\gamma}(0) = \eta(1, 0) = 0$ and $\widehat{\gamma}(1) = \eta(1, e) = e$ since $\varphi(0), \varphi(e) \leq a < b - 2\epsilon$). But, then, (2.3) and (2.4) above imply that

$$\max_{t\in[0,1]} \varphi(\widehat{\gamma}(t)) \leq c - \epsilon ,$$

which contradicts the definition of c. Therefore, c is a critical value of φ. □

Remark 2.1. In the case $u = 0$ is a strict local minimum of φ and $0 \neq e \in X$ is such that $\varphi(e) \leq \varphi(0)$, then Condition 2.1 is clearly satisfied. This situation is common in many application as we shall see next (in this sense, the rough Fig. 4.1 is typical).

3 Two Basic Applications

Application A. Let us show that the following nonlinear Dirichlet problem on a bounded domain $\Omega \subset \mathbb{R}^3$ with smooth boundary possesses a *classical* nontrivial solution:

$$\begin{cases} -\Delta u = u^3 & \text{in } \Omega \\ \quad u = 0 & \text{on } \partial\Omega . \end{cases} \tag{3.1}$$

To begin with, we observe that since $f(x,u) = u^3$ and $3 < \frac{N+2}{N-2} = 5$, the functional

$$\varphi(u) = \int_\Omega [\frac{1}{2}|\nabla u|^2 - \frac{1}{4}u^4]\ dx$$

is well defined and of class C^1 on the Sobolev space $H_0^1(\Omega)$ by Proposition 2.2.1. The critical points of φ are precisely the weak solutions of (3.1).

Lemma 3.1. *(a)* $u = 0$ *is a strict local minimum of* φ*;*
(b) Given $0 \neq v \in H_0^1$ *there exists* ρ_0 *such that* $\varphi(\rho_0 v) \leq 0$.

Proof: (a) In view of the Sobolev embedding $H_0^1 \subset L^4$ we have

$$\varphi(u) = \frac{1}{2}||u||^2 - \frac{1}{4}||u||_{L^4}^4 \geq \frac{1}{2}||u||^2 - C||u||^4\ ,$$

hence $\varphi(u) > 0 = \varphi(0)$ for all u with $0 < ||u|| \leq r$, for some small $r > 0$.
(b) Letting $\delta = \int_\Omega v^4 dx$ for a given $v \in H_0^1$ with (say) $||v|| = 1$, we have

$$\varphi(\rho v) = \frac{1}{2}\rho^2 - \frac{1}{4}\delta\rho^4 \to -\infty \quad \text{as } \rho \to \infty\ ,$$

so that the result follows. □

Theorem 3.2. *([9]) Problem* (3.1) *possesses a nontrivial classical solution.*[4]

Proof: We shall use the mountain-pass theorem. Since we already know that $\varphi \in C^1(H_0^1, \mathbb{R})$, we now show that φ satisfies (PS).

Let (u_n) be such that $|\varphi(u_n)| \leq C$, $\varphi'(u_n) \to 0$. Then, for all n sufficiently large, we have

$$|\varphi'(u_n) \cdot u_n| = |\int_\Omega [|\nabla u_n|^2 - u_n^4]\ dx| \leq ||u_n||\ ,$$

hence

$$\varphi(u_n) - \frac{1}{4}\varphi'(u_n) \cdot u_n\ dx \leq C + \frac{1}{4}||u_n||\ ,$$

that is,

$$\frac{1}{4}||u_n||^2 \leq C + \frac{1}{4}||u_n||\ .$$

This implies that $||u_n||$ is bounded, so that we may assume (by passing to a subsequence, if necessary) that $u_n \to \widehat{u}$ weakly in H_0^1. But then, since $\nabla\varphi(u) = u - T(u)$ with T a compact operator (cf. Remark 2.2.1), we obtain

[4] In fact, because of the *eveness* of the corresponding functional φ and its *superquadratic* nature, problem (3.1) has infinitely many solutions, as we shall see later on.

$$u_n = \nabla\varphi(u_n) + T(u_n) \longrightarrow 0 + T(\widehat{u}) .$$

Therefore, $u_n \to \widehat{u}$ strongly in H_0^1 and we have shown that φ satisfies (PS).

Now, Lemma 3.1 allows us to use Theorem 2.1 (with $e = \rho_0 v$) in order to conclude the existence of a critical point u_0 with $\varphi(u_0) = c \geq b > 0 = \varphi(0)$. Therefore, u_0 is a nontrivial weak solution of (3.1). Moreover, since both $\partial\Omega$ and $f(x,u) = u^3$ are smooth, a *bootstrap* argument shows that u_0 is indeed a classic solution (cf. [2]).

Application B. This next application is a generalization of the previous one. We consider the nonlinear Dirichlet problem (cf. [9])

$$\begin{cases} -\Delta u = f(x,u) & \text{in } \Omega \subset \mathbb{R}^N \\ \quad u = 0 & \text{on } \partial\Omega \, , \end{cases} \tag{3.2}$$

where $\Omega \subset \mathbb{R}^N$ $(N \geq 2)$ is a bounded smooth domain and, as usual, $f : \overline{\Omega} \times \mathbb{R} \longrightarrow \mathbb{R}$ is a Carathéodory function satisfying the growth condition (f_1) before Proposition 2.2.1 in Chapter 2. Moreover, we shall assume the following conditions:

$$f(x,s) = o(|s|) \;\text{ as } s \to 0, \text{ uniformly in } x. \tag{f_2}$$

There exist $\mu > 2$ and $r > 0$ such that

$$0 < \mu F(x,s) \leq s f(x,s) \quad \text{for } |s| \geq r, \tag{f_3}$$

uniformly in x (where we recall that $F(x,s) = \int_0^s f(x,\tau)d\tau$).

Condition (f_3) is the so-called superquadraticity condition of Ambrosetti and Rabinowitz.

As we know, the fact that f is a Carathéodory function satisfying (f_1) implies (cf. Proposition 2.2.1) that the functional

$$\varphi(u) = \int_\Omega [\frac{1}{2}|\nabla u|^2 - F(x,u)] \; dx \tag{3.3}$$

is well defined and is of class C^1 on the Sobolev space $H_0^1(\Omega)$. Next, we prove an analogue of Lemma 3.1.

Lemma 3.3. *(a) $u = 0$ is a strict local minimum of φ;*
(b) Given $0 \neq v \in H_0^1$ there exists ρ_0 such that $\varphi(\rho_0 v) \leq 0$.

Proof: *(a)* In view of (f_2), given $\epsilon > 0$, there exists $\delta = \delta(\epsilon) > 0$ such that $|f(x,s)| \leq \epsilon|s|$ for all $|s| \leq \delta$, hence

$$|F(x,s)| \le \frac{1}{2}\epsilon|s|^2 \text{ if } |s| \le \delta. \tag{3.4}$$

Now, since the growth condition (f_1) implies

$$|F(x,s)| \le A_\epsilon |s|^{\sigma+1} \quad \text{if } |s| \ge \delta = \delta(\epsilon)\ , \tag{3.5}$$

we combine (3.4) and (3.5) to get

$$|F(x,s)| \le \frac{1}{2}\epsilon|s|^2 + A_\epsilon|s|^{\sigma+1} \quad \forall s \in \mathbb{R}, \forall x \in \Omega. \tag{3.6}$$

Therefore, using (3.6) we obtain

$$\varphi(u) - \frac{1}{2}||u||^2 - \int_\Omega F(x,u)\ dx \ge \frac{1}{2}||u||^2 - \frac{\epsilon}{2}||u||^2_{L^2} - A_\epsilon||u||^{\sigma+1}_{L^{\sigma+1}}\ ,$$

hence

$$\varphi(u) \ge \frac{1}{2}||u||^2 - \frac{\epsilon}{2\lambda_1}||u||^2 - cA_\epsilon||u||^{\sigma+1} = \frac{1}{2}(1-\frac{\epsilon}{\lambda_1})||u||^2 - C_\epsilon||u||^{\sigma+1} \tag{3.7}$$

in view of Poincaré's inequality $\lambda_1||u||^2_{L^2} \le ||u||^2$ and the Sobolev inequality $||u||_{L^{\sigma+1}} \le c||u||$ (recall that $\sigma + 1 < \frac{2N}{N-2}$). Therefore, since we can take $\epsilon < \lambda_1$ and assume that $\sigma > 1$ in (f_1), the above inequality (3.7) gives $\varphi(u) > 0 = \varphi(0)$ for all u with $0 < ||u|| \le r$, for some suitably small $r > 0$.

(b) It is easy to see that condition (f_3), together with (f_1), implies that F is *superquadratic* in the sense that there exist constants $c, d > 0$ such that

$$F(x,s) \ge c|s|^\mu - d \quad \forall s \in \mathbb{R}, \forall x \in \Omega. \tag{3.8}$$

Therefore,

$$\varphi(u) = \frac{1}{2}||u||^2 - \int_\Omega F(x,u)\ dx \le \frac{1}{2}||u||^2 - c||u||^\mu_{L^\mu} + d|\Omega|\ ,$$

so that, given $v \in H^1_0$ with $||v|| = 1$ and writing $\delta = c||v||^\mu_{L^\mu} > 0$, we obtain

$$\varphi(\rho v) \le \frac{1}{2}\rho^2 - \delta\rho^\mu + d|\Omega| \longrightarrow -\infty \quad \text{as } \rho \to \infty.$$

In particular, there exists $\rho_0 > 0$ such that $\varphi(\rho_0 v) \le 0$. □

Remark 3.1. As we have just seen in part (b) of Lemma 3.3, condition (f_3) implies (3.8) with $\mu > 2$ (F is *superquadratic*) and, hence, $\varphi(\rho v) \to -\infty$ as $\rho \to \infty$ for any given $0 \ne v \in H^1_0$. Therefore, the functional φ is *not bounded* *from below.* On the other hand, since $\varphi(u) = \frac{1}{2}||u||^2 - \psi(u)$ where ψ is a weakly continuous functional (recall Example C in Section 2.1 of Chapter 2), then if we let (e_n) denote an orthonormal basis for H_0^1, it follows that $\lim_{n\to\infty} \psi(Re_n) = 0$ for any given $R > 0$, so that $\lim_{n\to\infty} \varphi(Re_n) = \frac{1}{2}R^2$. Since $R > 0$ is arbitrary, we see that φ is also *not bounded from above.*

Theorem 3.4. *([9]) If $f : \Omega \times \mathbb{R} \longrightarrow \mathbb{R}$ is a Carathéodory function satisfying conditions $(f_1) - (f_3)$, then problem (3.2) possesses a nontrivial weak solution $u \in H_0^1$.*

Proof: As in Theorem 3.2, we start by showing that the functional φ given in (3.3) satisfies the (PS) condition.

Let (u_n) be such that $|\varphi(u_n)| \leq C$, $\varphi'(u_n) \to 0$. Then, for all n sufficiently large, we have

$$|\varphi'(u_n) \cdot u_n| = |\int_\Omega [|\nabla u_n|^2 - f(x, u_n)u_n] \, dx| \leq ||u_n|| \; ,$$

hence

$$\varphi(u_n) - \frac{1}{\mu}\varphi'(u_n) \cdot u_n \; dx \leq C + \frac{1}{\mu}||u_n|| \; ,$$

that is,

$$(\frac{1}{2} - \frac{1}{\mu})||u_n||^2 \leq C + \frac{1}{\mu}||u_n|| \; ,$$

where $(\frac{1}{2} - \frac{1}{\mu}) > 0$, which implies that $||u_n||$ is bounded. The rest of the proof that φ satisfies (PS) is done as in Theorem 3.2. Similarly, Lemma 3.3 and Theorem 2.1 imply the existence of a nontrivial weak solution $u_0 \in H_0^1$ of (3.2). □

Remark 3.2. If $f : \overline{\Omega} \times \mathbb{R} \longrightarrow \mathbb{R}$ is assumed to be locally Lipschitzian, then by a *bootstrap* argument, the weak solution u_0 is a classical solution (see [9]).

Remark 3.3. We point out that the Palais–Smale condition is a compactness condition involving both the functional and the space X in a combined manner. The fact that X is infinite dimensional plays no role in requiring that (PS) (or some other compactness condition) be satisfied in the mountain-pass theorem. Indeed, even in a finite-dimensional space, the geometric conditions alone are not sufficient to guarantee that the level c is a critical level (see Exercise 2 that follows).

4 Exercises

1. Let $\lambda < 0$. Show that the ODE problem

$$\begin{cases} u'' + \lambda u + u^3 = 0 \,, & 0 < t < \pi \\ u'(0) = u'(\pi) = 0 \end{cases}$$

has a solution $u \in C^2[0,\pi]$ which is a mountain-pass critical point of the corresponding functional.

2. Find a polynomial function $p : \mathbb{R} \times \mathbb{R} \longrightarrow \mathbb{R}$ that satisfies the geometric conditions (2.1) of the mountain-pass theorem (so that the minimax value $c \geq b > 0$ does exist), but c is not a critical level of p. (Try to find such a polynomial $p(x,y)$ having $(0,0)$ as a strict local minimum and no other critical point; if giving up, see [20].)

3. Consider the following nonlinear Neumann problem

$$\begin{cases} -\Delta u \;=\; f(u) + \rho(x) & \text{in } \Omega \\ \dfrac{\partial u}{\partial n} = 0 \quad \text{on } \partial\Omega \,, \end{cases} \tag{N}$$

where $\Omega \subset \mathbb{R}^N$ $(N \geq 1)$ is a bounded smooth domain and the continuous functions $f : \mathbb{R} \longrightarrow \mathbb{R}$ (given as p-periodic) and $\rho : \overline{\Omega} \longrightarrow \mathbb{R}$ satisfy the conditions

$$\int_0^p f(s)\, ds = 0 \,, \qquad \int_\Omega \rho(x)\, dx = 0 \,.$$

Recall that, as an application of (the minimum principle) theorem 3.3.1 in Chapter 3 with the Palais–Smale condition replaced by the weaker Brézis–Coron–Nirenberg condition $(BCN)_c$, we proved that (N) had a solution $u_0 \in H^1(\Omega)$ minimizing the corresponding p-periodic functional φ. Clearly, by the periodicity of φ, any translated function $u_k = u_0 + kp$, $k \in \mathbb{Z}$, is also a minimizer of φ. Find another solution for (N) which is different from the u_k's.[5]

4. This is simply a calculus exercise to introduce a function which is *super-linear at infinity* in the sense that

$$\lim_{|s|\to\infty} \frac{f(s)}{s} = +\infty \,,$$

but grows *slower* than any power greater than 1, namely,

$$\lim_{|s|\to\infty} \frac{f(s)}{|s|^\epsilon s} = 0 \qquad \forall \epsilon > 0 \,.$$

[5] The mountain-pass theorem also holds if $b = a$ in (2.1) (cf. [63]).

Indeed, just take $f(s) := F'(s)$, where $F(s) = s^2 ln(1+s^2)$. You should also check that

$$\lim_{|s|\to\infty} [sf(s) - 2F(s)] = +\infty \,,$$

which is a condition that is relevant to the next exercise.

5. Consider the Dirichlet problem

$$\begin{cases} -\Delta u = f(x,u) & \text{in } \Omega \\ \quad u = 0 & \text{on } \partial\Omega \,, \end{cases} \qquad (D)$$

where $\Omega \subset \mathbb{R}^N$ is a bounded smooth domain and $f : \overline{\Omega} \times \mathbb{R} \longrightarrow \mathbb{R}$ is continuous, with $f(0) = 0$, $f(x,s) = o(|s|)$ as $s \to 0$ (uniformly for $x \in \Omega$), f satisfying the growth condition (f_1) in Chapter 2 and

$$\liminf_{|s|\to\infty} \frac{f(x,s)}{s} > \lambda_1 \,, \qquad \text{uniformly for } x \in \Omega \,,$$

Moreover, assume that

$$\lim_{|s|\to\infty} [sf(x,s) - 2F(x,s)] = +\infty \,, \qquad \text{uniformly for } x \in \Omega \,,$$ [6]

where, as usual, $F(x,s) = \int_0^s f(x,t)\,dt$. Show that (D) has a nonzero solution. [*Hint:* Use the Fatou lemma to verify that, in view of the above condition, the pertinent functional satisfies the Cerami condition introduced in Exercise 2 of Chapter 3.]

[6] This is a nonquadraticity condition introduced in [31].

5

The Saddle-Point Theorem

1 Preliminaries. The Topological Degree

In this chapter we present a second important illustration of the minimax method, the saddle-point theorem of Rabinowitz [64]. Since its proof uses the topological degree of Brouwer, we reserve this section for a brief presentation of this important topological tool and its main properties. The interested reader can see the details in [57].

Let $\Phi \in C(\overline{U}, \mathbb{R}^n)$ where $U \subset \mathbb{R}^n$ is a bounded open set. Given $b \in \mathbb{R}^n \backslash \Phi(\partial U)$, the problem consists in solving the equation

$$\Phi(x) = b \tag{1.1}$$

in U. This can be done in certain cases by using the so-called *Brouwer degree of the mapping* Φ (*with respect to* U, *at the point* b), denoted by $\deg(\Phi, U, b)$, which is an integer representing an *algebraic count* of the number of solutions of (1.1).

First, we consider the *regular case*, in which $\Phi \in C^1(\overline{U}, \mathbb{R}^n)$ and $b \in \mathbb{R}^n \backslash \Phi(\partial U)$ is a *regular value* of Φ [that is, $\Phi'(\xi)$ is invertible for any $\xi \in \Phi^{-1}(b)$], and define

$$\deg(\Phi, U, b) = \sum_{\xi \in \Phi^{-1}(b)} \operatorname{sgn} \det[\Phi'(\xi)] \,, \tag{1.2}$$

where we observe that the above sum is finite in view of the inverse function theorem.[1] Then the following properties hold:

[1] Indeed, the fact that $\Phi'(\xi)$ is invertible implies that each $\xi \in \Phi^{-1}(b)$ is isolated and, therefore, $\Phi^{-1}(b)$ is a finite set since $b \notin \Phi(\partial U)$ by assumption.

(i) (Normalization) If $Id : \overline{U} \longrightarrow \mathbb{R}^n$ is the inclusion mapping, then

$$deg(Id, U, b) = \begin{cases} 1 \; if \; b \in U \\ 0 \; if \; b \notin U \; . \end{cases}$$

(ii) (Existence Property) If $deg(\Phi, U, b) \neq 0$, then there exists a solution $x_0 \in U$ of (1.1).

(iii) (Additivity) If $U = U_1 \cup U_2$, $U_1 \cap U_2 = \emptyset$ and $b \notin \Phi(\partial U_1) \cup \Phi(\partial U_2)$, then

$$\deg(\Phi, U, b) = \deg(\Phi, U_1, b) + \deg(\Phi, U_2, b).$$

(iv) (Continuity) If Ψ is *close*[2] to Φ, then $\deg(\Psi, U, b) = \deg(\Phi, U, b)$.

(v) (Homotopy Invariance) If $H \in C([0,1] \times \overline{U}, \mathbb{R}^n)$ and $b \notin H([0,1] \times \partial U)$, then

$$\deg(H(t,\cdot), U, b) = \text{constant} \;\; \forall t \in [0,1] \; .$$

(vi) (Boundary Dependence) If $\Psi = \Phi$ on ∂U, then $\deg(\Psi, U, b) = \deg(\Phi, U, b)$.

Remark 1.1. It is not hard to verify properties (i)–(iv), whereas (v) and (vi) are consequences of (iv). In fact, (iv) implies that the function $t \mapsto deg(H(t,\cdot), U, b)$ is continuous, hence *constant* since it only takes integer values. And the homotopy $H(x,t) = (1-t)\Phi(x) + t\Psi(x)$ shows property (vi).

Next, the definition of $\deg(\Phi, U, b)$ can be extended to a general $\Phi \in C(\overline{U}, \mathbb{R}^n)$ and $b \in \mathbb{R}^n \backslash \Phi(\partial U)$ through the following steps:

(A) Given $\Phi \in C^1(\overline{U}, \mathbb{R}^n)$ and $b \in \mathbb{R}^n \backslash \Phi(\partial U)$, not necessarily a regular value, Sard's theorem implies the existence of a sequence (b_k) converging to b, where each b_k is a regular value of Φ; one defines $\deg(\Phi, U, b)$ by showing that the limit below exists and does not depend on the choice of (b_k):

$$\deg(\Phi, U, b) = \lim_{k \to \infty} \deg(\Phi, U, b_k) \; .$$

(B) Given $\Phi \in C(\overline{U}, \mathbb{R}^n)$ and $b \in \mathbb{R}^n \backslash \Phi(\partial U)$, we consider a sequence (Φ_k) where $\Phi_k \in C^1(\overline{U}, \mathbb{R}^n)$ is such that $\Phi_k \to \Phi$ in $C(\overline{U}, \mathbb{R}^n)$ and we define $deg(\Phi, U, b)$ by showing that the limit below exists and does not depend on the choice of (Φ_k):

$$\deg(\Phi, U, b) = \lim_{k \to \infty} \deg(\Phi_k, U, b) \; .$$

[2] Here we mean C^1-*close* but, in the general case to be considered, we should understand *close* as C^0-*close*.

Finally, it follows that the *degree* just defined satisfies properties (i)-(vi). We should also mention that there is a degree theory in the infinite-dimensional case, due to Leray and Schauder (see [50, 57]) which deals with mappings $\Phi \in C(\overline{U}, X)$, where $U \subset X$ is a bounded open subset of a Banach space X and Φ is a *compact perturbation* of the identity, that is, of the form $\Phi(u) = u - T(u)$ with T a compact mapping. The corresponding *Leray–Schauder degree* also satisfies properties (i)–(vi).

2 The Abstract Result

In this section we state and prove Rabinowitz's saddle-point theorem.

Theorem 2.1. *([64]) Let $X = V \oplus W$ be a Banach space, where $\dim V < \infty$, and let $\varphi \in C^1(X, \mathbb{R})$ be a functional satisfying the Palais–Smale condition (PS). If D is a bounded neighborhood of 0 in V such that*

$$a := \max_{\partial D} \varphi < \inf_{W} \varphi := b \;, \tag{2.1}$$

then

$$c := \inf_{h \in \Gamma} \max_{u \in \overline{D}} \varphi(h(u))$$

is a critical value of φ with $c \geq b$. (Here Γ is the class of deformations *of $\overline{D}$ in X which fix ∂D pointwise, that is, $\Gamma = \{h \in C(D, X) \,|\, h(u) = u, \;\; \forall u \in \partial D \;\}$.)*

Proof: We first verify that $h(\overline{D}) \cap W \neq \emptyset$ for any $h \in \Gamma$. In fact, if we let $P : X \longrightarrow V$ denote the projection onto V along W, then $Ph \in C(\overline{D}, V)$ and $Ph(u) = Pu = u \neq 0 \; \forall u \in \partial D$. Since we can identify V with $\mathbb{R}^n$, where $n = \dim V$, the degree $deg(Ph, D, 0)$ is well defined and properties (i) and (vi) of Section 1 imply that

$$\deg(Ph, D, 0) = \deg(Id, D, 0) = 1.$$

Therefore, by property (ii) of that section, there exists $u_0 \in D$ such that $Ph(u_0) = 0$, that is, $h(u_0) \in W$. This implies that

$$\max_{u \in \overline{D}} \varphi(h(u)) \geq b = \inf_{W} \varphi \;,$$

hence $c \geq b$ since $h \in \Gamma$ is arbitrary.

Now, assume by negation that c is not a critical value of φ. Then, by the deformation theorem 3.2.2, there exist $0 < \epsilon < \frac{b-a}{2}$ (recall that $a < b$ by (2.1)) and $\eta \in C([0,1] \times X, X)$ such that

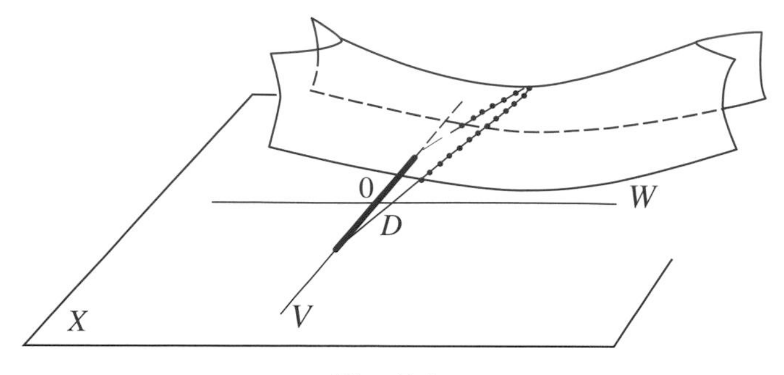

Fig. 5.1.

$$\eta(t,u) = u \quad if \quad u \notin \varphi^{-1}([c-2\epsilon, c+2\epsilon]) \ , \quad t \in [0,1] \ , \tag{2.2}$$

$$\eta(1, \varphi^{c+\epsilon}) \subset \varphi^{c-\epsilon} \ . \tag{2.3}$$

Now, pick $h \in \Gamma$ such that

$$\max_{u \in \overline{D}} \varphi(h(u)) \leq c + \epsilon \tag{2.4}$$

and define $\widehat{h}(u) = \eta(1, h(u))$. By (2.2) and the fact that $2\epsilon < b - a$ we have $\eta(1, h(u)) = u$ if $u \in \partial D$ [since $\varphi|\partial D \leq a < b - 2\epsilon$], hence $\widehat{h} \in \Gamma$. Then, (2.3) and (2.4) imply that

$$\max_{u \in \overline{D}} \varphi(\widehat{h}(u)) \leq c - \epsilon \ ,$$

which contradicts the definition of c. Therefore, c is a critical value of φ. □

Remark 2.1. We should observe that Condition (2.1) is clearly satisfied when φ is such that $\varphi(v) \to -\infty$ as $||v|| \to \infty$, $v \in V$, and $\varphi(w) \to +\infty$ as $||w|| \to \infty$, $w \in W$. This situation is typical in many applications.

3 Application to a Resonant Problem

We now consider a result due to Ahmad, Lazer and Paul [3] concerning the existence of a solution for the following *resonant problem*

$$\begin{cases} -\Delta u = \lambda_k u + g(x,u) & \text{in } \Omega \subset \mathbb{R}^N \\ \quad u = 0 \quad \text{on } \partial\Omega \ , \end{cases} \tag{3.1}$$

where $\Omega \subset \mathbb{R}^N$ $(N \geq 1)$ is a bounded smooth domain, λ_k is the kth eigenvalue of $-\Delta u = \lambda u$ in Ω, $u = 0$ on $\partial\Omega$, and we assume

$g : \overline{\Omega} \times \mathbb{R} \longrightarrow \mathbb{R}$ is continuous and uniformly bounded, say

$$|g(x,s)| \leq M \quad \forall x \in \Omega, \quad \forall s \in \mathbb{R} \ . \tag{g_1}$$

Therefore the nonlinearity $f(x,s) = \lambda_k s + g(x,s)$ is such that $\frac{f(x,s)}{s} \to \lambda_k$ as $|s| \to \infty$, hence the terminology *resonant problem.* Clearly, some additional condition is necessary for the solvability of (3.1) since, as we know, even in the linear case $g(x,s) = g(x)$ (say, a continuous function), one has the *Fredholm alternative*: (3.1) is solvable if and only if $\int_\Omega gv dx = 0$ for all v in the λ_k-eigenspace N_k.

Ahmad, Lazer and Paul [3] assume that g also satisfies one of the conditions (g_2^+) or (g_2^-) below:

$$\int_\Omega G(x, v(x))\, dx \to \pm\infty \quad \text{as } ||v|| \to \infty,\ v \in N_k \ , \tag{$g_2^\pm$}$$

where, as usual, $G(x,s) = \int_0^s g(x,t)dt$. More precisely, they prove the following:

Theorem 3.1. *([3]) Under conditions (g_1) and (g_2^+) or (g_2^-) Problem (3.1) has a weak solution $u \in H_0^1(\Omega)$.*

Here we present the proof given by Rabinowitz [64] which uses the saddle-point theorem and makes clear the role of conditions $(g_2^\pm)$[3]. For that, we consider the functional

$$\varphi(u) = \int_\Omega [\frac{1}{2}(|\nabla u|^2 - \lambda_k u^2) - G(x,u)]\, dx = \frac{1}{2}\langle Lu, u\rangle - \int_\Omega G(x,u)\, dx \ , \tag{3.2}$$

which, in view of (g_1), is clearly well defined on $H_0^1(\Omega)$, is of class C^1, and its critical points are the weak solutions of (3.1). We also consider the following orthogonal decomposition of $X = H_0^1$,

$$X = X_- \oplus X_0 \oplus X_+ \ ,$$

where $X_0 = N_k$ and X_+ (resp. X_-) is the subspace where the operator $L : H_0^1 \to H_0^1$ defined in (3.2) is positive (resp. negative) definite. We denote the respective orthogonal projections by P_0, P_+ and P_-. Then, the following result holds true:

Proposition 3.2. *Under conditions (g_1) and (g_2^-) one has*

[3] In fact, it was this result of Ahmad, Lazer and Paul that led Rabinowitz to state and prove the saddle-point theorem, which became one additional powerful tool among the variational techniques.

(a) $\varphi(u) \to -\infty$ *as* $||u|| \to \infty$, $u \in X_-$;
(b) $\varphi(u) \to +\infty$ *as* $||u|| \to \infty$, $u \in X_0 \oplus X_+$.

Proof: *(a)* Let $u = P_- u \in X_-$. Using (g_1) together with the mean value theorem for $G(x, \cdot)$ and the fact that L is negative definite on X_-, we obtain

$$\begin{aligned}\varphi(u) &= \frac{1}{2}\langle LP_- u, P_- u\rangle - \int_\Omega G(x, P_- u)\, dx \\ &\le -\frac{1}{2}\alpha ||P_- u||^2 + M||P_- u||_{L^1} \\ &\le -\frac{1}{2}\alpha ||P_- u||^2 + A||P_- u|| \longrightarrow -\infty \quad \text{as } ||u|| = ||P_- u|| \to \infty,\end{aligned}$$

where, in the last inequality, we used that $||v||_{L^1} \le c||v||_{L^2}$ and the Poincaré inequality.

(b) Let $u = P_0 u + P_+ u \in X_0 \oplus X_+$. Now, using the fact that L is positive definite on X_+ and, again, (g_1) combined with the mean value theorem for $G(x, \cdot)$, we obtain

$$\begin{aligned}\varphi(u) &= \frac{1}{2}\langle LP_+ u, P_+ u\rangle - \int_\Omega G(x, u)\, dx \\ &\ge \frac{1}{2}\alpha ||P_+ u||^2 - \int_\Omega [G(x, u) - G(x, P_0 u)]\, dx - \int_\Omega G(x, P_0 u)\, dx \\ &\ge \frac{1}{2}\alpha ||P_+ u||^2 - M||P_+ u||_{L^1} - \int_\Omega G(x, P_0 u)\, dx\ ,\end{aligned}$$

hence

$$\varphi(u) \ge \frac{1}{2}\alpha ||P_+ u||^2 - A||P_+ u|| - \int_\Omega G(x, P_0 u)\, dx \tag{3.3}$$

for all $u = P_0 u + P_+ u \in X_0 \oplus X_+$. This shows, in view of hypothesis (g_2^-), that $\varphi(u) \to +\infty$ as $||u||^2 = ||P_0 u||^2 + ||P_+ u||^2 \to \infty$. □

Remark 3.1. If we assumed (g_2^+) instead of (g_2^-) the conclusions of the above proposition would be

(c) $\varphi(u) \to -\infty$ as $||u|| \to \infty$, $u \in X_- \oplus X_0$;
(d) $\varphi(u) \to +\infty$ as $||u|| \to \infty$, $u \in X_+$.

Proof of Theorem 3.1 As mentioned earlier, we are going to use the saddle-point theorem. Since we already know that $\varphi \in C^1(X, \mathbb{R})$, we show now that φ satisfies (PS).

Let (u_n) be such that $|\varphi(u_n)| \leq C$ and $\varphi'(u_n) \to 0$. Then, for all n sufficiently large, we have

$$|\varphi'(u_n) \cdot h| = |\langle Lu_n, h\rangle - \int_\Omega g(x, u_n)h \, dx| \leq ||h|| \quad \forall h \in X. \tag{3.4}$$

Then, letting $h = P_+u_n$ in (3.4), it follows that

$$\begin{aligned} ||P_+u_n|| &\geq |\varphi'(u_n) \cdot (P_+u_n)| \\ &\geq \alpha||P_+u_n||^2 - M||P_+u_n||_{L^1} \\ &\geq \alpha||P_+u_n||^2 - A||P_+u_n|| \ , \end{aligned}$$

so that $||P_+u_n||$ must be bounded.

Next, letting $h = P_-u_n$ in (3.4), we obtain

$$\begin{aligned} -||P_-u_n|| &\leq \varphi'(u_n) \cdot (P_-u_n) \\ &\leq -\alpha||P_-u_n||^2 + M||P_-u_n||_{L^1} \\ &\leq -\alpha||P_-u_n||^2 + A||P_-u_n|| \ , \end{aligned}$$

so that $||P_-u_n||$ must also be bounded. Therefore, we obtain

$$||(P_+ + P_-)u_n|| = ||u_n - P_0u_n|| \leq C \ . \tag{3.5}$$

On the other hand, since we can write

$$\begin{aligned} \varphi(u_n) = &\frac{1}{2}\langle L(u_n - P_0u_n), u_n - P_0u_n\rangle \\ &- \int_\Omega [G(x, u) - G(x, P_0u)] \, dx - \int_\Omega G(x, P_0u) \, dx \ , \end{aligned}$$

where $\varphi(u_n)$ is bounded together with the first two terms on the right-hand side of the above expression (in view of (3.5)), it follows that the last term is also bounded and, hence, because of (g_2^-), we can conclude that $||P_0u_n||$ is bounded. This, together with (3.5) shows that $||u_n||$ is bounded. The rest of the proof that φ satisfies (PS) is done as before, using the fact that $\nabla\varphi(u) = u - T(u)$ with T a compact operator.

Finally, Proposition 3.2 allows us to apply the saddle-point theorem with $V = X_-$, $W = X_0 \oplus X_+$ to conclude the existence of a critical point of φ, that is, a solution of (3.1). □

Remark 3.2. We point out that in the case $g : \overline{\Omega} \times \mathbb{R} \longrightarrow \mathbb{R}$ verifies

$$\lim_{|s|\to\infty} G(x,s) = +\infty \quad \text{uniformly for} \quad x \in \overline{\Omega}\,, \tag{g_3^+}$$

respectively

$$\lim_{|s|\to\infty} G(x,s) = -\infty \quad \text{uniformly for} \quad x \in \overline{\Omega}\,, \tag{g_3^-}$$

then condition (g_2^+) [resp. (g_2^-)] is satisfied.

Indeed, following [63], let $v \in N_k$ and $K > 0$ be given and write $v = \rho\omega$ with $\rho = ||v||$ and $\omega \in S \cap N_k = \{\omega \in N_k \mid ||\omega|| = 1\}$. By (g_3^+) there exists $\alpha = \alpha(K)$ such that $G(x,s) \geq K$ if $|s| \geq \alpha(K)$ and $x \in \overline{\Omega}$. Then, since there exists M_0 such that $G(x,s) \geq -M_0$ for all $x \in \overline{\Omega}$ and $s \in \mathbb{R}$, we can write

$$\int_\Omega G(x, \rho\omega(x))\, dx \geq -M_0|\Omega| + K|\Omega_{\rho,K,\omega}| \tag{3.6}$$

where $\Omega_{\rho,K,\omega} = \{x \in \Omega \mid \rho|w(x)| \geq \alpha(K)\}$. Since $S \cap N_k$ is a compact set, we can choose $\rho(K)$ sufficiently large so that $|\Omega_{\rho,K,\omega}| \geq |\Omega|/2$ for all $\rho \geq \rho(K)$ and $\omega \in S \cap N_k$. Thus, (3.6) implies that (g_2^+) is satisfied.

4 Exercises

1. As can be easily checked by elementary methods, the linear *resonant* ODE problem (P_2) in the introduction has no solution. Why cannot Theorem 3.1 be used?
2. In view of the Fredholm alternative, if $\rho : \overline{\Omega} \longrightarrow \mathbb{R}$ is a continuous function on a bounded, smooth domain $\Omega \subset \mathbb{R}^N$, a necessary and sufficient condition for existence of a solution to the linear problem $-\Delta u = \lambda_1 u + \rho(x)$, $u = 0$ on $\partial\Omega$, is that
$$\int_\Omega \rho(x)\phi_1(x)\, dx = 0\,,$$
where $\phi_1(x)$ is an eigenfunction corresponding to λ_1. Since both conditions (g_2^+) and (g_2^-) clearly fail in this linear case, exhibit a class of continuous functions $g : \mathbb{R} \longrightarrow \mathbb{R}$ for which the modified nonlinear problem with $\rho(x)g(s)$ replacing $\rho(x)$ always has a solution.
3. Let $\Omega \subset \mathbb{R}^N$ $(N \geq 1)$ be a bounded, smooth domain and consider the Dirichlet problem

$$\begin{cases} -\Delta u = \lambda_1 u + g(u) - h(x) & \text{in } \Omega \\ u = 0 & \text{on } \partial\Omega\,, \end{cases} \tag{D}$$

where $h : \overline{\Omega} \longrightarrow \mathbb{R}$ and $g : \mathbb{R} \longrightarrow \mathbb{R}$ are continuous functions with g increasing and such that

$$g(-\infty) < g(s) < g(+\infty) \quad \forall s \in \mathbb{R},$$

where $g(\pm\infty) := \lim_{s\to\pm\infty} g(s)$. Show that, if $\phi_1 > 0$ is a λ_1-eigenfunction and the following Landesman–Lazer condition [49]

$$g(-\infty)\int_\Omega \phi_1\,dx < \int_\Omega h(x)\phi_1\,dx < g(+\infty)\int_\Omega \phi_1\,dx \qquad (LL)$$

is satisfied, then problem (D) has a solution. Verify that the condition (LL) is also necessary for existence of a solution under the other given conditions.

4. State a corresponding Neumann problem where existence of a solution can be proved in a manner similarly to the one used in the previous exercise.

5. Show that the functional $\varphi(u)$ associated with the Dirichlet problem in Exercise 3 satisfies

$$\varphi(t\phi_1) \longrightarrow -\infty \qquad \text{as } |t| \to \infty\,.$$

6. Consider the nonlinear Dirichlet problem

$$\begin{cases} -\Delta u = f(x,u) & \text{in } \Omega \\ \quad u = 0 & \text{on } \partial\Omega\,, \end{cases} \qquad (D)$$

where $\Omega \subset \mathbb{R}^N$ is a bounded smooth domain and $f : \overline{\Omega}\times\mathbb{R} \longrightarrow \mathbb{R}$ is a continuous function such that

$$\lim_{|s|\to\infty} \frac{f(x,s)}{s} = \lambda\,, \qquad \text{uniformly for } x \in \Omega\,,$$

where $\lambda_k < \lambda < \lambda_{k+1}$ (λ_k, λ_{k+1} being eigenvalues of $-\Delta$ under Dirichlet condition on $\partial\Omega$). Assume further that

$$\lim_{|s|\to\infty} [2F(x,s) - sf(x,s)] = +\infty\,, \qquad \text{uniformly for } x \in \Omega\,,$$

where, as usual, $F(x,s) = \int_0^s f(x,t)\,dt$. Show that (D) has a solution. (Recall that, as in Exercise 5 of Chapter 4, the previous condition renders the pertinent functional satisfying the Cerami condition.)

6

Critical Points under Constraints

1 Introduction. The Basic Minimization Principle Revisited

In many variational problems one must find critical points of a given functional $\varphi \in C^1(X, \mathbb{R})$ in the presence of *constraints*, that is, critical points of φ restricted to a set $M \subset X$ of constraints. Naturally, in order to be able to talk about critical points of $\varphi|M$, the set M must have a *differentiable structure*. Typically, in the case of a finite number of constraints, M is of the form $M = \{u \in X \mid \psi_j(u) = 0 \ , \ \ j = 1, \dots, k\}$ where $\psi_j \in C^1(X, \mathbb{R})$, $j = 1, \dots, k$.

Before considering critical points in general, let us review the situation of *minimum* described in Section 2.1, where the set M of constraints does not need to have a differentiable structure. Then, the same ideas of the basic minimization principle, Theorem 2.1.3, can be used to prove the following:

Theorem 1.1. *Let M be a* weakly closed *subset of a Hilbert space (or, more generally, a reflexive Banach space) X. Suppose a functional $\varphi : M \longrightarrow \mathbb{R}$ is*

(*i*) weakly lower semicontinuous (w.l.s.c),
(*ii*) coercive, that is, $\varphi(u) \to +\infty$ *as* $||u|| \to \infty$, $u \in M$.

Then φ is bounded from below and there exists $u_0 \in M$ such that $\varphi(u_0) = \inf_M \varphi$.

Proof: The proof is essentially the same as that of Theorem 2.1.3, where we replaced $\overline{B}_R$ by the bounded, weakly closed set $M_R = M \cap \overline{B}_R$ and recalled that a *bounded, weakly closed* set is *weakly compact.* And, as before, in view of the coerciveness of φ, we pick $p \in M$ and $R > 0$ so that $\varphi(u) \geq \varphi(p)$ for $||u|| \geq R$, $u \in M$. □

Remark 1.1. The special case $M = X$ gives us Theorem 2.1.2. In this case, as already observed in the remark following Theorem 2.1.2, any point u_0 of minimum of a C^1 functional $\varphi : X \longrightarrow \mathbb{R}$ is a critical point of φ, that is, $\varphi'(u_0) = 0 \in X^*$.

Remark 1.2. A typical example of a *weakly closed* set $M \subset X$ is given by $M = \{u \in X \mid \psi(u) = c\}$, where $\psi : X \longrightarrow \mathbb{R}$ is a *weakly continuous* functional. More generally, if $\psi : X \longrightarrow \mathbb{R}$ is a weakly l.s.c. functional, the sets of the form $\psi^c = \{u \in X \mid \varphi(u) \leq c\}$, $c \in \mathbb{R}$, are weakly closed.

2 Natural Constraints

Now, we turn our attention to the problem of finding the critical points of a functional $\varphi \in C^1(X, \mathbb{R})$ over a set of constraints $M \subset X$ which is a differentiable manifold. Besides the trivial case where M is an arbitrary closed subspace of X, we will not consider a general submanifold $M \subset X$, but rather those submanifolds of *finite codimension.* These correspond to situations in which only a finite number of constraints are present. Let us recall some definitions.

A subset $M \subset X$ is a *C^m-submanifold of codimension* k ($m, k \geq 1$ integers) if, for each $u_0 \in M$, there exists an open neighborhood U of u_0 and a function $\psi \in C^m(U, \mathbb{R}^k)$ such that

(i) $\psi'(u)$ is *surjective* for every $u \in U$,
(ii) $M \cap U = \{u \in U \mid \psi(u) = 0\}$.

In this case the *tangent space to M at the point $u_0 \in M$*, denoted by $T_{u_0}M$, is defined as the space of all *tangent vectors* $\gamma'(0)$ to M at u_0, where $\gamma : (-\epsilon, \epsilon) \longrightarrow M$ is an arbitrary C^1 curve in M passing through u_0, that is, $\gamma(0) = u_0$, and such that $\gamma'(t) \neq 0$ $\forall t \in (-\epsilon, \epsilon)$.

Finally, given a functional $\varphi \in C^1(X, \mathbb{R})$, we say a point $u_0 \in M$ is a *critical point of* $\varphi|M$ if

$$\frac{d}{dt}\varphi(\gamma(t))|_{t=0} = 0 \tag{2.1}$$

for every C^1 curve $\gamma : (-\epsilon, \epsilon) \longrightarrow M$ passing through u_0.

The following result, which is well known in finite dimension (and whose proof we omit in infinite dimension, cf. Theorem 3.1.31 in [15]), shows that a critical point of a *constrained functional* is a critical point of a related *unconstrained functional.* (This is the so-called *method of Lagrange multipliers.*)

Theorem 2.1. *Let $\varphi \in C^1(X, \mathbb{R})$ and suppose $M \subset X$ is a C^1-submanifold of codimension k, say $M = \{u \in U \mid \psi_j(u) = 0,\ j = 1, \dots, k\}$ where $\psi_j \in C^1(U, \mathbb{R})$, $j = 1, \dots, k$, $U \subset X$ is an open set and $\psi_1'(u), \dots, \psi_k'(u)$ are linearly independent for each $u \in U$. Then, if $\widehat{u} \in M$ is a critical point of $\varphi|M$, there exists $\widehat{\lambda} = (\lambda_1, \dots, \lambda_k) \in \mathbb{R}^k$ such that*

$$\varphi'(\widehat{u}) = \widehat{\lambda} \cdot \psi'(\widehat{u}) = \sum_{j=1}^{k} \lambda_j \psi_j'(\widehat{u}) \ . \tag{2.2}$$

(*The λ_j's are called* Lagrange multipliers.)

Next, given a submanifold $M \subset X$ as above, if $u_0 \in M$ is a critical point of an unconstrained functional $\varphi \in C^1(X, \mathbb{R})$, then u_0 is also a critical point of $\varphi|M$ since (2.1) is easily verified in this case ($\varphi'(u_0) = 0 \in X^*$ implies, in particular, that $\varphi'(u_0) \cdot \gamma'(0) = 0$ for all tangent vectors $\gamma'(0)$ to M at u_0). The converse statement is not true in general. However, if a given critical point $\widehat{u} \in M$ of $\varphi|M$ is also a critical point of the unconstrained functional φ, we say that the submanifold M is a *natural constraint for the critical point* $\widehat{u} \in M$ (in view of (2.2), this is the case if and only if all the Lagrange multipliers $\lambda_j = \lambda_j(\widehat{u})$ are equal to zero). We say that M is a *natural constraint for* φ if M is a natural constraint for each critical point $\widehat{u}$ of $\varphi|M$, that is,

$$K(\varphi|M) = K(\varphi) \cap M \tag{2.3}$$

(cf. [45]), where we are denoting by $K(\widetilde{\varphi})$ the set of critical points of $\widetilde{\varphi}$.

A particularly important situation occurs when a given critical point u_0 of $\varphi \in C^1(X, \mathbb{R})$ is *not* a *global minimum*, say, a *saddle-point*[1], but it is possible to find a submanifold $M \subset X$ which is a *natural constraint* for φ and is such that $u_0 \in M$ is a *global minimum* for $\varphi|M$ (cf. the definition of natural constraint in [15], Section 3.3B, where in addition M is required to contain *all* critical points of φ, that is, $K(\varphi|M) = K(\varphi)$). We shall see some applications in the next section.

[1] This means that any neighborhood of u_0 contains points u, v with $\varphi(u) < \varphi(u_0) < \varphi(v)$.

3 Applications

Application A. We consider the question of existence of T-periodic solutions for the classic Hamiltonian system

$$\begin{cases} \ddot{x} + \nabla_x V(t,x) = 0 \\ x(0) = x(T) \ , \quad \dot{x}(0) = \dot{x}(T) \ , \end{cases} \tag{3.1}$$

where $x(t) \in \mathbb{R}^n$ and $V(t,x)$ satisfies

$V : \mathbb{R} \times \mathbb{R}^n \longrightarrow \mathbb{R}$ is continuous, $V(\cdot, x)$ is T-periodic and $V(t,\cdot)$ is differentiable with $\nabla_x V(t,x)$ continuous for any $t \in \mathbb{R}$, $x \in \mathbb{R}^n$. (V_1)

Let X denote the following Sobolev space of T-periodic functions, $X = H_T^1 = \{x : \mathbb{R} \to \mathbb{R}^n \mid x \text{ is abs. continuous, } T\text{-periodic and } \int_0^T |\dot{x}|^2 dt < \infty \}$, endowed with its natural inner product

$$\langle x, \widehat{x} \rangle = \int_0^T [x \cdot \widehat{x} + \dot{x} \cdot \dot{\widehat{x}}] \, dt \quad \forall x, \widehat{x} \in X \ ,$$

where $\cdot$ and $|\cdot|$ denote the Euclidean scalar product and norm in $\mathbb{R}^n$.

Lemma 3.1. *If* (V_1) *holds, then the functional*

$$\varphi(x) = \int_0^T [\frac{1}{2}|\dot{x}|^2 - V(t,x)] \, dt \ , \quad x \in X = H_T^1 \tag{3.2}$$

is well defined and of class C^1 *on* X*, with*

$$\varphi'(x) \cdot z = \int_0^T [\dot{x} \cdot \dot{z} - \nabla_x V(t,x) \cdot z] \, dt \quad \forall x, z \in X \ . \tag{3.3}$$

Moreover, the critical points of φ *are the classical* C^2 *solutions of* (3.1).

Proof: Since V is continuous on $\mathbb{R} \times \mathbb{R}^n$ by (V_1) and any function $x \in X$ is absolutely continuous with $\int_0^T |\dot{x}|^2 dt < \infty$, it is clear that the functional φ is well defined on X. Also, the fact that $\varphi \in C^1(X, \mathbb{R})$ with the derivative given by (3.3) can be easily verified, so that we leave these as an exercise for the reader (cf. Exercise 2).

Finally, the proof that the critical points of φ are the classic C^2 solutions of (3.1) is also standard. Indeed, if $x \in X$ is a critical point of φ, then

$$\int_0^T [\dot{x} \cdot \dot{z} - \nabla_x V(t,x) \cdot z] \, dt = 0 \quad \forall z \in X \ ,$$

which implies that $x \in X$ is a weak solution of (3.1). Since $x \in X$ is continuous and, by (V_1), $\nabla_x V$ is also continuous, it follows that $\ddot{x} = v$ in the weak sense, where $v(t) = \nabla_x V(t, x(t))$ is also a continuous function. Therefore, the weak solutions $x(t)$ are classical C^2 solutions (cf. [70]). □

Remark 3.1. It should be noted that, in this case, one also has the gradient of φ, $\nabla\varphi : X \longrightarrow X$, of the form $\nabla\varphi(x) = x - K(x)$ with $K : X \longrightarrow X$ a compact operator. In fact, since $K(x)$ is defined by

$$\langle K(x), z\rangle = \int_0^T [x + \nabla_x V(t,x)] \cdot z \, dt \quad \forall x, z \in X \ ,$$

the compactness of K in this case is an immediate consequence of the compact embedding $X \subset C[0,T]$ (see Exercise 3).

Lemma 3.2. *(cf. [76]) Assume the conditions (V_1) and (V_2), where*

$$\nabla_x V(-t, -x) = -\nabla_x V(t,x) \quad \forall t, x \qquad (V_2)$$

(i.e., $\nabla_x V$ is an odd *function of t, x). Then, the closed subspace of the* odd *functions in X,*

$$M = \{x \in X \mid x(-t) = -x(t) \ \forall t\}$$

is a natural constraint *for φ.*

Proof: Let $x_0 \in M$ be a critical point of $\varphi|M$, that is,

$$\langle \nabla\varphi(x_0), z\rangle = \int_0^T [\dot{x}_0 \cdot \dot{z} - \nabla_x V(t, x_0) \cdot z] \, dt = 0 \quad \forall z \in M \ .$$

Since we can decompose $X = M \oplus M^\perp$, where $M^\perp$ is the subspace of the *even* functions in X, in order to show that $x_0 \in M$ is a critical point of φ, it suffices to verify that

$$\langle \nabla\varphi(x_0), w\rangle = \int_0^T [\dot{x}_0 \cdot \dot{w} - \nabla_x V(t, x_0) \cdot w] \, dt = 0 \quad \forall w \in M^\perp \ .$$

But this is true since (V_2) and the fact that x_0 is an *odd* function imply that the above integrand is an *odd* (T-periodic) function. □

Lemma 3.3. *(Wirtinger Inequality) If $x \in X$ and $\int_0^T x \, dt = 0$, then*

$$\int_0^T |x|^2 \, dt \le \frac{T^2}{4\pi^2} \int_0^T |\dot{x}|^2 \, dt \ .$$

Proof: It is a consequence of the Parseval identity applied to the functions x and $\dot{x}$, by noticing that the *constant term* in the Fourier expansion of the function $x \in X$ is *zero*, since x has zero-mean value. □

Remark 3.2. Note that Wirtinger inequality is the analogue in the periodic situation of the Poincaré inequality.

Theorem 3.4. *([76])Assume conditions* $(V_1) - (V_3)$, *where*

There exist constants $0 \le a < \frac{4\pi^2}{T^2}$ *and* $b \ge 0$ *such that*

$$V(t,x) \le \frac{1}{2}a|x|^2 + b\ . \tag{V_3}$$

Then, Problem (3.1) *has a classical,* odd *solution, which minimizes the functional* φ *over* $M = \{x \in X \mid x \text{ is odd }\}$. *If, in addition,* V *satisfies*

$$[\nabla_x V(t,x) - \nabla_x V(t,\widehat{x})] \cdot [x - \widehat{x}] < \frac{4\pi^2}{T^2}|x - \widehat{x}|^2 \quad \forall x, \widehat{x} \in \mathbb{R}^n, \quad x \ne \widehat{x}, \tag{V_4}$$

then (3.1) *has a* unique (*classical, odd*) *solution.*

Proof: Since M is a natural constraint by Lemma 3.2, it suffices to find a critical point of $\varphi|M$.

If $x \in M$, we have $\int_0^T x dt = 0$, so that Wirtinger's inequality (Lemma 3.3) implies the following lower estimate for φ, where $(1 - \frac{aT^2}{4\pi^2}) > 0$:

$$\varphi(x) \ge \frac{1}{2}\int_0^T |\dot{x}|^2\, dt - \frac{a}{2}\int_0^T |x|^2\, dt - bT$$
$$\ge \frac{1}{2}(1 - \frac{aT^2}{4\pi^2})\int_0^T |\dot{x}|^2\, dt - bT \quad \forall x \in M.$$

Therefore, φ is coercive on M. Since φ is weakly l.s.c. (being the sum of a continuous convex functional and a weakly continuous functional), Theorem 1.1 implies the existence of $x_0 \in M$ such that $\varphi(x_0) = \inf_M \varphi$.

Clearly $x_0 \in M$ is a critical point of $\varphi|M$, hence a critical point of φ (since M is a natural constraint) and a classical solution of (3.1) by Lemma 3.1.

Finally, if (V_4) is valid, then using the Wirtinger inequality yields

$$\langle \nabla\varphi(x) - \nabla\varphi(\widehat{x}), x - \widehat{x}\rangle$$
$$= \int_0^T (|\dot{x} - \dot{\widehat{x}}|^2 - [\nabla_x V(t,x) - \nabla_x V(t,\widehat{x})] \cdot [x - \widehat{x}])\, dt$$
$$> \int_0^T (|\dot{x} - \dot{\widehat{x}}|^2 - \frac{4\pi^2}{T^2}|x - \widehat{x}|^2)\, dt \ge 0$$

for any $x, \widehat{x} \in M$, $x \neq \widehat{x}$, so that we have uniqueness of the solution in this case. □

Application B. Consider the following Neumann problem

$$\begin{cases} -\Delta u = f(u) & \text{in } \Omega \\ \frac{\partial u}{\partial n} = 0 & \text{on } \partial\Omega \,, \end{cases} \tag{3.4}$$

where $\Omega \subset \mathbb{R}^N$ $(N \geq 1)$ is a bounded smooth domain and $f : \mathbb{R} \longrightarrow \mathbb{R}$ is C^1, strictly increasing, such that $f(0) = 0$ and the limits $f_{\pm} = \lim_{s \to \pm\infty} f'(s)$ exist and satisfy

$$0 < f_{\pm} < \lambda_1 < f'(0) \,, \tag{3.5}$$

where λ_1 is the first positive eigenvalue of the problem $-\Delta u = \lambda u$ in Ω, $\frac{\partial u}{\partial n} = 0$ on $\partial\Omega$.

It is clear that (3.4) has the trivial solution $u(x) = 0$. Here, using the ideas in [45, 16], we will show that (3.4) possesses a nontrivial solution which can be found by minimizing the functional

$$\varphi(u) = \int_{\Omega} [\frac{1}{2}|\nabla u|^2 - F(u)] \, dx \,, \quad \text{where } F(s) = \textstyle\int_0^s f \text{ as usual} \,, \tag{3.6}$$

over a suitable submanifold M of $H^1(\Omega)$ which is a *natural constraint for* φ. Such a solution, which is of a *saddle-point type* for the above functional, can be found by other methods, e.g., Amann's reduction method [4] or, since F is convex, by the duality method of Clarke and Ekeland [26] (we will treat this latter method in the next chapter).

Consider the subset M of $X = H^1(\Omega)$ defined by

$$M = \{u \in X \mid \int_{\Omega} f(u) \, dx = 0 \} \,.$$

Lemma 3.5. *Under condition* (3.5) *one has*

(*i*) *$M \subset X$ is a C^1 submanifold of codimension* 1*;*

(*ii*) *$u \in X$ is a critical point of φ if and only if $u \in M$ and u is a critical point of $\varphi|M$.*

Proof: (*i*) In view of (3.5) and the fact that f is strictly increasing, we have $0 < f'(s) < C$ for all $s \in \mathbb{R}$ and some constant $C > 0$. Therefore, the functional

$$\psi(u) = \int_{\Omega} f(u) \, dx$$

is of class C^1 on X with Frechét derivative $\psi' : X \longrightarrow X^*$ given by

$$\psi'(u) \cdot h = \int_\Omega f'(u)h \, dx \quad \forall h \in X,$$

and satisfying $\psi'(u) \neq 0$ for all $u \in X$.

(ii) If $u \in X$ is a critical point of φ, then

$$\varphi'(u) \cdot h = \int_\Omega [\nabla u \cdot \nabla h - f(u)h] \, dx = 0 \quad \forall h \in X \ ,$$

so that, by choosing $h = 1 \in X$, we get $\int_\Omega f(u)dx = 0$, i.e., $u \in M$, and thus u is a critical point of $\varphi|M$. Conversely, if $u \in M$ is critical point of $\varphi|M$, then by Theorem 2.1, there exists $\lambda \in \mathbb{R}$ such that

$$\int_\Omega [\nabla u \cdot \nabla h - f(u)h] \, dx = \lambda \int_\Omega f'(u)h \, dx \quad \forall h \in X.$$

Therefore, by choosing $h = 1$ again and using the fact that $\int_\Omega f(u)dx = 0$, we obtain $\lambda = 0$ (since f is strictly increasing). This shows that u is a critical point of φ. □

Remark 3.3. The previous lemma not only shows that M is a *natural constraint for* φ but, in fact, shows that M contains *all* critical points of φ (i.e., M is a natural constraint according to [15, Section 6.3B].

Next, similarly to what we did in Section 3.3, we decompose the Sobolev space $X = H^1(\Omega)$ as

$$X = X_0 \oplus X_1 \ , \tag{3.7}$$

where $X_1 = \mathbb{R} = \text{span}\,\{1\}$ is the (one-dimensional) space of the constant functions on Ω and $X_0 = (\text{span}\,\{1\})^\perp = \{v \in X \mid \int_\Omega v \, dx = 0 \}$ is the space of functions in $H^1(\Omega)$ having mean value equal to zero. Also, we recall that the following Poincaré inequality holds for a function in X_0:

$$\lambda_1 \int_\Omega v^2 \, dx \leq \int_\Omega |\nabla v|^2 \, dx \quad \forall v \in X_0. \tag{3.8}$$

Lemma 3.6. *Under Condition* (3.5), *and writing $u \in M$ as $u = v + c$, where $v \in X_0$, $c \in X_1 = \mathbb{R}$, one has*

(i) $\int_\Omega F(v + c) \, dx \leq \int_\Omega F(v) \, dx \quad \forall v + c \in M;$

(ii) $||v_n|| \to \infty$ *as* $||v_n + c_n|| \to \infty$, $v_n + c_n \in M$.

Proof: (*i*) The convexity of F implies the inequality

$$F(s) - F(\widehat{s}) \geq f(\widehat{s})(s - \widehat{s}) \quad \forall s, \widehat{s} \in \mathbb{R}.$$

Letting $s = v(x)$, $\widehat{s} = v(x) + c$ and integrating over Ω gives

$$\int_\Omega F(v)\, dx - \int_\Omega F(v + c)\, dx \geq \int_\Omega f(v + c)(-c)\, dx = 0 \,,$$

since $v + c \in M$.

(*ii*) Define the function $g : X_0 \times \mathbb{R} \longrightarrow \mathbb{R}$ by the formula

$$g(v, c) = \int_\Omega f(v + c)\, dx \,.$$

(In other words, $g(v, c) = \psi(v + c)$ in the notation of Lemma 3.5 (i) and $g(v, c) = 0$ if and only if $v + c \in M$.) Then, for each $v \in X_0$ fixed, the function $g(v, \cdot)$ is strictly increasing since f is also. In fact, since $a \leq f'(s) \leq b$ for all $s \in \mathbb{R}$ (and some $0 < a < b$), it is not hard to see that, for each $v \in X_0$, there exists a unique $c = c(v) \in \mathbb{R}$ such that $v + c \in M$.

Now, assume by contradiction that there exists a sequence $v_n + c_n \in M$ with $||v_n + c_n|| \to \infty$ and $||v_n||$ bounded. Then, we may assume that there exists $\widehat{v} \in X_0$ such that $v_n \rightharpoonup \widehat{v}$ weakly in X, $v_n \to \widehat{v}$ strongly in $L^1(\Omega)$ and $|c_n| \to \infty$, say $c_n \to \infty$. By picking $\alpha > c(\widehat{v})$ and an integer n_0 such that $g(\widehat{v}, c_n) > g(\widehat{v}, \alpha)$ for all $n \geq n_0$, we obtain

$$0 < g(\widehat{v}, \alpha) \leq g(\widehat{v}, c_n) = \int_\Omega [f(\widehat{v} + c_n) - f(v_n + c_n)]\, dx \leq b \int_\Omega |\widehat{v} - v_n|\, dx \,,$$

which is absurd since the last integral approaches zero. □

Theorem 3.7. *Under Condition* (3.5), *Problem* (3.4) *possesses a nontrivial weak solution* $u_0 \in H^1(\Omega)$ *minimizing the functional* φ *over the manifold* $M = \{u \in H^1(\Omega) \mid \int_\Omega f(u) dx = 0 \}$.

Proof: We can use Theorem 1.1 by noticing that M is weakly closed since the functional $\psi : H^1(\Omega) \longrightarrow \mathbb{R}$, $\psi(u) = \int_\Omega f(u) dx$, is weakly continuous. Also, the functional

$$\varphi(u) = \int_\Omega [\frac{1}{2}|\nabla u|^2 - F(u)]\, dx \quad u \in H^1(\Omega),$$

is weakly l.s.c., since it is the sum of a continuous, convex functional and a weakly continuous functional (Recall Examples B and C in Section 2.1). Therefore, it remains to prove that φ is coercive on M.

For that, we first observe that (3.5) implies the existence of constants $0 < \beta < \lambda_1$ and $\gamma \in \mathbb{R}$ such that

$$F(s) \leq \frac{1}{2}\beta s^2 + \gamma \quad \forall s \in \mathbb{R}. \tag{3.9}$$

Then, using Lemma 3.6 above we obtain, for $v + c \in M$,

$$\begin{aligned}\varphi(v+c) &= \int_\Omega [\frac{1}{2}|\nabla v|^2 - F(v+c)]\,dx \geq \int_\Omega [\frac{1}{2}|\nabla v|^2 - F(v)]\,dx \\ &\geq \frac{1}{2}\int_\Omega |\nabla v|^2\,dx - \frac{1}{2}\beta\int_\Omega v^2\,dx - \gamma|\Omega|\end{aligned}$$

and, thus, in view of the Poincaré inequality (3.8),

$$\varphi(v+c) \geq \frac{1}{2}(1 - \frac{\beta}{\lambda_1})\int_\Omega |\nabla v|^2\,dx - \gamma|\Omega|\,, \tag{3.10}$$

where $(1 - \frac{\beta}{\lambda_1}) > 0$. Now, if $||v + c|| \to \infty$, $v + c \in M$, then Lemma 3.6 (ii) implies that $||v|| \to \infty$ or, equivalently (by Poincaré's inequality), that $\int_\Omega |\nabla v|^2 dx \to \infty$. But, then, (3.10) shows that $\varphi(v + c) \to \infty$, so that φ is coercive on M.

In view of Theorem 1.1 there exists $u_0 \in M$ such that $\varphi(u_0) = \inf_M \varphi$. In order to show that $u_0 \neq 0$, it suffices to find $\widehat{u} \in M$ such that $\varphi(\widehat{u}) < 0 = \varphi(0)$. For that, we use the other half of Condition (3.5), namely $\lambda_1 < f'(0)$, to conclude the existence of constants $\mu > \lambda_1$ and $\rho > 0$ such that

$$F(s) \geq \frac{1}{2}\mu s^2 \quad \text{if } |s| \leq \rho. \tag{3.11}$$

Then, by choosing $\widehat{u} = \epsilon v_1 + c = \widehat{v} + c \in M$ where v_1 is a λ_1-eigenfunction and $\epsilon > 0$ and $c > 0$ are sufficiently small so that $||\widehat{u}||_{L^\infty(\Omega)} \leq \rho$, we use (3.11) to obtain

$$\begin{aligned}\varphi(\widehat{u}) &= \frac{1}{2}\lambda_1 \int_\Omega \widehat{v}^2\,dx - \int_\Omega F(\widehat{u})\,dx \\ &\leq \frac{1}{2}(\lambda_1 - \mu)\int_\Omega \widehat{v}^2\,dx - \frac{1}{2}\mu\int_\Omega [(\widehat{v}+c)^2 - \widehat{v}^2]\,dx \\ &= \frac{1}{2}(\lambda_1 - \mu)\int_\Omega \widehat{v}^2\,dx - \frac{1}{2}\mu(\int_\Omega 2c\widehat{v}\,dx + c^2|\Omega|) \\ &\leq \frac{1}{2}(\lambda_1 - \mu)\int_\Omega \widehat{v}^2\,dx\,,\end{aligned}$$

where we recall that $\int_\Omega \widehat{v}dx = 0$ since $\widehat{v} \in X_0$. □

Application C. Next, we consider the following periodic problem

$$\begin{cases} \ddot{u} + k(t)e^u = h(t) \\ u(0) = u(2\pi) \quad , \quad \dot{u}(0) = \dot{u}(2\pi) \ , \end{cases} \tag{3.12}$$

where $h, k : \mathbb{R} \longrightarrow \mathbb{R}$ are continuous, 2π-periodic functions, with[2]

$$\int_0^{2\pi} h(t) \, dt = 0 \ . \tag{3.13}$$

This situation is the 1-dimensional analogue (which we chose for simplicity) of the problem of finding *conformally equivalent* Riemannian structures on compact manifolds with prescribed Gauss curvature (see [15, Sections 1.1A and 6.4B]). The problem here is to find necessary and sufficient conditions on the periodic function $k(t)$ for problem (3.12) to have a solution.

In what follows, we denote by X the Sobolev space $X = H^1_{2\pi} = \{u : \mathbb{R} \longrightarrow \mathbb{R} \mid u$ is absolutely continuous, $2\pi - \textit{periodic}$ and $\int_0^{2\pi} \dot{u}^2 \, dt < \infty\}$ with its natural inner product $\langle \cdot, \cdot \rangle$, and consider the *quadratic* functional

$$\varphi(u) = \int_0^{2\pi} [\frac{1}{2}\dot{u}^2 + hu] \, dt \ ,$$

which, as easily seen, is of class C^∞ on X, with derivative

$$\varphi'(u) \cdot v = \int_0^{2\pi} [\dot{u}\dot{v} + hv] \, dt \quad \forall v \in X \ .$$

In fact, we observe that the critical points of φ are the solutions of the linear problem

$$\begin{cases} \ddot{u} = h(t) \\ u(0) = u(2\pi) \quad , \quad \dot{u}(0) = \dot{u}(2\pi) \ , \end{cases} \tag{3.14}$$

which are given by the formula

$$u(t) = \left(\frac{1}{2t} \int_0^{2\pi} sh(s) \, ds \right) t + \int_0^t (t-s)h(s) \, ds + c = u_0(t) + c \tag{3.15}$$

where $c \in \mathbb{R}$ is an arbitrary constant. Let us consider the following condition on the function k,

$$\int_0^{2\pi} k(t)e^{u_0(t)} \, dt < 0 \ , \tag{3.16}$$

where $u_0(t)$ is defined in (3.15), and let us define the following subset $M \subset X$:

$$M = \{u \in X \mid \int_0^{2\pi} ke^u \, dt = 0 \ , \ \int_0^{2\pi} u \, dt = 0 \ \} \ .$$

[2] Note the necessity of condition (3.13) for the solvability of (3.12) when $k(t) = 0$.

Lemma 3.8. *Assume conditions* (3.13) *and* (3.16). *Then*

(*i*) *$M \subset X$ is a C^1 submanifold of codimension 2;*
(*ii*) *Any critical point of $\varphi|M$ is (up to a constant) a solution of Problem* (3.12).

Proof: (*i*) It is not hard to see (cf. Exercise 6) that $M \neq \emptyset$ and that the functionals $\psi_1, \psi_2 : X \longrightarrow \mathbb{R}$ defined by

$$\psi_1(u) = \int_0^{2\pi} ke^u \, dt \quad , \quad \psi_2(u) = \int_0^{2\pi} u \, dt \ ,$$

are of class C^1 (in fact, C^∞). Moreover, $\psi_1'(u)$ and $\psi_2'(u)$ are linearly independent for any $u \in M$. Indeed, if

$$[\alpha\psi_1'(u) + \beta\psi_2'(u)] \cdot v = \alpha \int_0^{2\pi} ke^u v \, dt + \beta \int_0^{2\pi} v \, dt = 0 \quad \forall v \in X \ ,$$

we obtain $\beta = 0$ (by choosing $v = 1$) and, then, also $\alpha = 0$ (by choosing $v = e^{u_0 - u}$ and using (3.16)). Therefore, $M \subset X$ is a C^1 submanifold of codimension 2.

(*ii*) Let $u \in M$ be a critical point of $\varphi|M$. Then, there exist $\lambda_1, \lambda_2 \in \mathbb{R}$ such that

$$\int_0^{2\pi} [\dot{u}\dot{v} + hv] \, dt = \lambda_1 \int_0^{2\pi} ke^u v \, dt + \lambda_2 \int_0^{2\pi} v \, dt = 0 \quad \forall v \in X \ ,$$

which implies that $u \in X$ is a weak solution, hence a classical solution of the problem $-\ddot{u} + h = \lambda_1 ke^u + \lambda_2$, $u(0) = u(2\pi)$, $\dot{u}(0) = \dot{u}(2\pi)$. By integrating this last equation over $[0, 2\pi]$ and using (3.13) together with the fact that $\int_0^{2\pi} ke^u \, dt = 0$ (as $u \in M$), we obtain $\lambda_2 = 0$. It necessarily follows that $\lambda_1 \neq 0$ since, otherwise, u would be a solution of (3.14), that is, of the form $u = u_0 + c$, but, on the other hand, the fact that $\int_0^{2\pi} ke^u \, dt = 0$ (as $u \in M$) contradicts (3.16). We have shown that a critical point $u \in M$ of $\varphi|M$ is a solution of the problem

$$-\ddot{u} + h = \lambda_1 ke^u \ , \quad u(0) = u(2\pi) \quad , \quad \dot{u}(0) = \dot{u}(2\pi) \ , \tag{3.17}$$

where $\lambda_1 \neq 0$. If we prove that $\lambda_1 > 0$, then, by letting $e^c = \lambda_1$ and defining $\widehat{u} = u + c$, we can verify that

$$\ddot{\hat{u}} + ke^{\hat{u}} = \ddot{u} + \lambda_1 ke^u = h \ , \quad \hat{u}(0) = \hat{u}(2\pi) \quad , \quad \dot{\hat{u}}(0) = \dot{\hat{u}}(2\pi) \ .$$

In other words, $\widehat{u} = u + c$ is a solution of Problem (3.12) as we wanted to show. Finally, in order to prove that $\lambda_1 > 0$, we write $u = u_0 + w$ in (3.17), multiply the resulting equation by e^{-w} and integrate over $[0, 2\pi]$ to obtain (since u_0 is a solution of (3.14)) that

$$-\int_0^{2\pi} e^{-w}\dot{w}^2 \, dt = \lambda_1 \int_0^{2\pi} ke^{u_0} \, dt \, .$$

Thus, (3.16) implies that $\lambda_1 > 0$. □

Remark 3.4. The previous lemma shows that the manifold M is kind of a *natural constraint* for the functional φ.

Theorem 3.9. *Suppose that* (3.13) *holds true and* $k(t) \neq 0$. *Then, Problem* (3.12) *has a solutions if and only if* $k(t)$ *changes sign and* (3.16) *is satisfied.*

Proof: (i) If u is a solution of problem (3.12), then by integrating the corresponding equation over $[0, 2\pi]$, we obtain $\int_0^{2\pi} ke^u \, dt = 0$, so that $k(t)$ must change sign. Moreover, writing $u = u_0 + w$, multiplying the resulting equation by e^{-w} and integrating over $[0, 2\pi]$, we obtain (again, recalling that $\ddot{u}_0 = h$)

$$\int_0^{2\pi} e^{-w}\dot{w}^2 \, dt + \int_0^{2\pi} ke^{u_0} \, dt = 0 \, ,$$

and, since $w \neq$ *constant* (as $k(t) \neq 0$), we conclude that (3.16) is necessarily satisfied.

(ii) Conversely, suppose that (3.16) is satisfied. Then, in view of Lemma 3.8, in order to show that Problem (3.12) has a solution, it suffices to find a critical point of $\varphi|M$. For that, we observe that the functional φ is coercive on M since $M \subset X_0 = \{u \in X \mid \int_0^{2\pi} u \, dt = 0\}$ and the Wirtinger inequality holds true in X_0 (cf. Lemma 3.3). Indeed, we have

$$\varphi(u) = \int_0^{2\pi} [\frac{1}{2}\dot{u}^2 + hu] \, dt \geq \frac{1}{2}||u||^2 - ||h||_{L^2}||u|| \longrightarrow +\infty$$

as $||u|| \to \infty$, $u \in X_0$.

Therefore, since φ is weakly l.s.c. and M is weakly closed in X, Theorem 1.1 shows the existence of $u_0 \in M$ such that $\varphi(u_0) = \inf_M \varphi$, so that $u_0 \in M$ is a critical point of $\varphi|M$. □

4 Exercises

1. Let $\varphi \in C^1(X, \mathbb{R})$, where X is a Hilbert space, and let V be a closed subspace of X. Suppose that $u_0 \in V$ is a critical point of the constrained functional $\varphi|V$. If V is invariant under $\nabla\varphi$ (i.e., $\nabla\varphi(V) \subset V$), verify that u_0 is a critical point of the unconstrained functional φ. (This result is related to the *principle of symmetric criticality* of Palais [61], where the set Fix (G) of *fixed elements of X under the action of a group G* (see Chapter 8) is one such invariant subspace V.)

2. Show that $\varphi \in C^1(X, \mathbb{R})$, where φ is a functional satisfying (V_1) as defined in Lemma 3.1

3. Suppose $x : [0,1] \longrightarrow \mathbb{R}^n$ is an absolutely continuous function, that is, an antiderivative of an L^1-function $v : [0,1] \longrightarrow \mathbb{R}^n$. If, in addition, one assumes that $v \in L^2([0,1], \mathbb{R}^n)$ (i.e., $v \in H^1([0,1], \mathbb{R}^n)$), show that the embedding $H^1([0,1], \mathbb{R}^n) \subset C([0,1], \mathbb{R}^n)$ is compact. Use this fact to fill in the details of Remark 3.1.

4. Consider the following nonlinear Dirichlet problem

$$\begin{cases} -\Delta u = f(u) & \text{in } \Omega \\ \quad u = 0 & \text{on } \partial\Omega\,, \end{cases} \tag{P}$$

where $\Omega \subset \mathbb{R}^N$ $(N \geq 1)$ is a bounded domain and $f : \mathbb{R} \longrightarrow \mathbb{R}$ is a C^1-function satisfying the subcritical growth condition

$$|f'(s)| \leq c|s|^{p-2} + d\,,$$

for some $c, d > 0$ and $2 \leq p < 2N/(N-2)$ if $N \geq 3$ $(2 \leq p < \infty$ if $N = 1, 2)$. Let $X = H_0^1(\Omega)$ and $\varphi : X \longrightarrow \mathbb{R}$ be the functional associated with problem (P). If, in addition, the function f satisfies the condition

$$f'(s) > \frac{f(s)}{s} \qquad \forall\, s \neq 0\,,$$

show that the set $M = \{\, u \in X \backslash \{0\} \mid \int_\Omega [|\nabla u|^2 - f(u)u]\, dx = 0 \,\} \subset X$ is a C^1-submanifold of codimension 1, which is a natural constraint for φ.

5. Provide the details in the proof of Lemma 3.3 and verify that the constant in the Wirtinger inequality is sharp.

6. Verify that $M \neq \emptyset$ in Lemma 3.8 and that the functionals ψ_1, ψ_2 considered in that lemma are of class C^∞.

7

A Duality Principle

1 Convex Functions. The Legendre–Fenchel Transform

A function $F : \mathbb{R}^n \longrightarrow \mathbb{R}$ is said to be *convex* if, for any $u, v \in \mathbb{R}^n$ and $\lambda \in [0,1]$, one has

$$F((1-\lambda)u + \lambda v) \leq (1-\lambda)F(u) + \lambda F(v) .$$

The function F is called strictly convex if one has strict inequality when $u \neq v$ and $\lambda \neq 0, 1$. It is easy to verify that if $F \in C^1(\mathbb{R}^n, \mathbb{R})$, then F is convex if and only if the gradient of F, $\nabla F : \mathbb{R}^n \longrightarrow \mathbb{R}^n$, is a *monotone mapping*, that is,

$$(\nabla F(u) - \nabla F(v), u - v) \geq 0 \quad \forall u, v \in \mathbb{R}^n ,$$

where we are denoting by $(\cdot, \cdot)$ the usual inner product in $\mathbb{R}^n$. Moreover, $F \in C^1(\mathbb{R}^n, \mathbb{R})$ is *strictly convex* if and only if ∇F is *strictly monotone* (i.e., the above inequality is strict whenever $u \neq v$).

Proposition 1.1. *Let $F \in C^1(\mathbb{R}^n, \mathbb{R})$ be convex and such that $\nabla F : \mathbb{R}^n \longrightarrow \mathbb{R}^n$ is $1-1$ and onto. Then, for each $v \in \mathbb{R}^n$, the function $u \mapsto (v, u) - F(u)$ is bounded from above and the formula*

$$G(v) = \sup_u [(u, v) - F(u)] \tag{1.1}$$

defines a function $G : \mathbb{R}^n \longrightarrow \mathbb{R}$ such that $G \in C^1(\mathbb{R}^n, \mathbb{R})$ and $\nabla G = (\nabla F)^{-1}$.

Remark 1.1. $G := F^*$ is called the *Legendre–Fenchel transform* of F. The assumptions in the above Proposition 1.1 are satisfied when, for example, $F \in C^1(\mathbb{R}^n, \mathbb{R})$ is strictly convex and $F(u)/|u| \to \infty$ as $|u| \to \infty$. Indeed, since F is strictly convex, ∇F is clearly $1-1$. On the other hand, for each fixed $v \in \mathbb{R}^n$, the function $u \mapsto F(u) - (v,u)$ is coercive in view of the coercivity of $F(u)/|u|$. Therefore, the function $u \mapsto F(u) - (v,u)$ is bounded from below and there exists $\widehat{u} \in \mathbb{R}^n$ such that $F(\widehat{u}) - (v, \widehat{u}) = \inf_u [F(u) - (v,u)]$. Since $F \in C^1(\mathbb{R}^n, \mathbb{R})$, we necessarily have $\nabla F(\widehat{u}) - v = 0$, that is, $\nabla F(\widehat{u}) = v$. Thus, ∇F is also onto $\mathbb{R}^n$.

Proof of Proposition 1.1: First, since F is of class C^1 and convex, we observe that

$$F(w) \geq F(u) + (\nabla F(u), w - u) \tag{1.2}$$

for any $u, w \in \mathbb{R}^n$. Indeed, the convexity of F implies that

$$\frac{1}{\lambda}[F(u + \lambda(w-u)) - F(u)] \leq F(w) - F(u)$$

for $\lambda \in [0,1]$, so that, passing to the limit as $\lambda \to 0$ yields (1.2).

Now, if for any given $v \in \mathbb{R}^n$ we define $u := (\nabla F)^{-1}(v)$ and use (1.2), we obtain

$$(v,w) - F(w) \leq (v,u) - F(u) \quad \forall w \in \mathbb{R}^n.$$

Therefore, the function $w \mapsto (v,w) - F(w)$ is bounded from above and its supremum is attained when $w = u = (\nabla F)^{-1}(v)$, that is,

$$G(v) = \sup_w [(v,w) - F(w)] = (v, (\nabla F)^{-1}(v)) - F((\nabla F)^{-1}(v)) \,. \tag{1.3}$$

In particular, one obtains the so-called *Fenchel inequality*,

$$(v,w) \leq F(w) + G(v) \quad \forall v, w \in \mathbb{R}^n. \tag{1.4}$$

In order to show that $G \in C^1(\mathbb{R}^n, \mathbb{R})$ with $\nabla G = (\nabla F)^{-1}$, we fix $v_0 \in \mathbb{R}^n$ and write $u_0 = (\nabla F)^{-1}(v_0)$, $u_h = (\nabla F)^{-1}(v_0 + h)$ to obtain, from (1.3),

$$G(v_0 + h) - G(v_0) = (h, u_h) + (v_0, u_h - u_0) - F(u_h) + F(u_0) \,.$$

Hence

$$G(v_0 + h) - G(v_0) \leq (h, u_h) \tag{1.5}$$

in view of (1.2). On the other hand, using (1.4) with $v = v_0 + h$, $w = u_0$ gives

$$G(v_0 + h) - G(v_0) \geq (v_0 + h, u_0) - F(u_0) - G(v_0) \,,$$

hence

$$G(v_0 + h) - G(v_0) \geq (h, u_0) , \tag{1.6}$$

since $G(v_0) = (v_0, u_0) - F(u_0)$. Combining (1.5) and (1.6) we obtain

$$0 \leq G(v_0 + h) - G(v_0) - (h, u_0) \leq (h, u_h - u_0) ,$$

which implies $G(v_0 + h) - G(v_0) - (h, u_0) = o(h)$ as $|h| \to 0$, since $u_h - u_0 = (\nabla F)^{-1}(v_0 + h) - (\nabla F)^{-1}(v_0)$ and $(\nabla F)^{-1} : \mathbb{R}^n \longrightarrow \mathbb{R}^n$ is continuous by the *theorem on domain invariance* (cf. [15, Theorem 5.4.11]. We have shown that there exists $\nabla G(v_0)$ and that $\nabla G(v_0) = u_0 = (\nabla F)^{-1}(v_0)$. Moreover, $\nabla G = (\nabla F)^{-1}$ is continuous, hence $G \in C^1(\mathbb{R}^n, \mathbb{R})$. □

Typical Example. If $F(u) = \frac{1}{p}||u|^p$, with $p > 1$, then $G(v) = \frac{1}{q}|v|^q$, where $q = \frac{p}{p-1}$ is the *conjugate exponent* of p. In particular, $F^* = F$ if $F(u) = \frac{1}{2}|u|^2$.

Proposition 1.2. *Let $F_1, F_2 \in C^1(\mathbb{R}^n, \mathbb{R})$ be convex functions such that $\nabla F_1, \nabla F_2 : \mathbb{R}^n \longrightarrow \mathbb{R}^n$ are $1-1$ and onto. If $F_1 \leq F_2$, then $F_1^* \geq F_2^*$. In addition, if there exist constants $\beta > 0$ and $\delta \in \mathbb{R}$ such that*

$$F(u) \leq \beta \frac{|u|^2}{2} + \delta \quad \forall u \in \mathbb{R}^n , \tag{1.7}$$

then

$$F^*(v) \geq \frac{1}{\beta}\frac{|v|^2}{2} - \delta \quad \forall v \in \mathbb{R}^n . \tag{1.8}$$

(Similarly, the above implications also hold with the inequalities $\leq$ and $\geq$ exchanged.)

Proof: It is immediate from (1.1) that $F_1 \leq F_2$ implies $G_1 \geq G_2$. Also, note that $F_\beta^*(v) = \frac{1}{2\beta}|v|^2$ if $F_\beta(u) = \frac{\beta}{2}|u|^2$, so that we obtain (1.8) from (1.7). □

Remark 1.2. Let $F \in C^1(\mathbb{R}^n, \mathbb{R})$ be strictly convex and satisfy

$$\alpha \frac{|u|^2}{2} - \delta \leq F(u) \leq \beta \frac{|u|^2}{2} + \delta \quad \forall u \in \mathbb{R}^n , \tag{1.9}$$

where $0 < \alpha \leq \beta$ and $\delta \in \mathbb{R}$. Then, $F(u)/|u| \to \infty$ as $|u| \to \infty$ and F^* is well defined by Remark 1.1. Moreover, Proposition 1.2 and (1.9) above imply that

$$\frac{1}{\beta}\frac{|v|^2}{2} - \delta \leq F^*(v) \leq \frac{1}{\alpha}\frac{|v|^2}{2} + \delta \quad \forall v \in \mathbb{R}^n . \tag{1.10}$$

2 A Variational Formulation for a Class of Problems

Let $L : D(L) \subset X \longrightarrow X$ be a selfadjoint operator on the Hilbert space $X = L^2(\Omega; \mathbb{R}^n)$, where $\Omega \subset \mathbb{R}^N$ is a bounded domain. Also, let $F \in C^1(\mathbb{R}^n, \mathbb{R})$ be a function satisfying the condition

$$|\nabla F(u)| \leq a|u| + b \quad \forall u \in \mathbb{R}^n ,$$

with $a, b \geq 0$, so that the Nemytskii operator $\nabla F : X \longrightarrow X$ is well defined and continuous. Then, the problem

$$Lu + \nabla F(u) = 0 \quad , \quad u \in D(L) , \tag{P}$$

has the following *variational formulation*:
Find the critical points of the functional

$$\varphi(u) = \frac{1}{2}(Lu, u)_X + \int_\Omega F(u)\, dx \ , \quad u \in D(L) . \tag{ϕ}$$

Here, in view of the closed graph theorem, we are considering $D(L) = Z$ as a Hilbert space when endowed with the *graph inner product*, $(u_1, u_2)_Z = (u_1, u_2)_X + (Lu_1, Lu_2)_X$, $\forall u_1, u_2 \in Z$. In fact, we have the following

Proposition 2.1. *The functional $\varphi : Z \longrightarrow \mathbb{R}$ given by (ϕ) is of class C^1 and*

$$\varphi'(u) \cdot h = (Lu + \nabla F(u), h)_X \quad \forall u, h \in Z . \tag{2.1}$$

In particular, $u \in D(L) = Z$ is a solution of (P) if and only if u is a critical point of φ.

Proof: Let $u, h \in Z$. Then

$$\begin{aligned}\delta(u;h) &= \varphi(u+h) - \varphi(u) - (Lu + \nabla F(u), h)_X \\ &= (Lu, h)_X + \frac{1}{2}(Lh, h)_X + \int_\Omega F(u+h)\, dx \\ &\quad - \int_\Omega F(u)\, dx - (Lu + \nabla F(u), h)_X ,\end{aligned}$$

that is,

$$\begin{aligned}\delta(u;h) &= \int_\Omega F(u+h)\, dx - \int_\Omega F(u)\, dx - (\nabla F(u), h)_X + \frac{1}{2}(Lh, h)_X \\ &= \gamma(u;h) + \frac{1}{2}(Lh, h)_X ,\end{aligned}$$

where $|\gamma(u;h)|/||h||_X \to 0$ as $||h||_X \to 0$, since, as we know, the functional $\psi : u \mapsto \int_\Omega F(u)dx$ is of class C^1 on X with $\psi'(u) \cdot h = (\nabla F(u), h)_X$. Therefore, since $||h||_Z \geq ||h||_X \ \forall h \in Z$, we obtain, *a fortiori* that $|\gamma(u;h)|/||h||_Z \to 0$ as $||h||_Z \to 0$. Also, it is clear that $|(Lh,h)_X|/||h||_Z \to 0$ as $||h||_Z \to 0$, $h \in Z$, since $|(Lh,h)_X| \leq ||Lh||_X||h||_X \leq ||h||_Z^2$. Therefore,

$$|\delta(u;h)|/||h||_Z \to 0$$

as $||h||_Z \to 0$, $h \in Z$, which shows that the Fréchet derivative φ' exists and is given by (2.1). Moreover, formula (2.1) gives the continuity of the mapping $Z \ni u \mapsto \varphi'(u) \in Z^*$. Indeed, we have

$$\begin{aligned} ||\varphi'(u+v) - \varphi'(u)||_{Z^*} &\leq \sup_{||h||_Z \leq 1} ||Lv + \nabla F(u+v) - \nabla F(u)||_X ||h||_X \\ &\leq ||Lv||_X + ||\nabla F(u+v) - \nabla F(u)||_X \to 0 \end{aligned}$$

as $||v||_Z \to 0$, $v \in Z$ (again, we use the continuity of $X \ni u \mapsto \psi'(u) \in X^*$).□

3 A Dual Variational Formulation

In this section we introduce a variational formulation which is *dual* to that considered in the previous section, and which constitutes a *variant* of the *duality principle* due to Clarke and Ekeland for Hamiltonian systems (cf. [26]). Our presentation will follow the work [17] of Brézis (see also [30]).

As before, $L : D(L) \subset X \longrightarrow X$ is a selfadjoint operator on the Hilbert space $X = L^2(\Omega;\mathbb{R}^n)$, $\Omega \subset \mathbb{R}^N$ a bounded domain, and $F : \mathbb{R}^n \longrightarrow \mathbb{R}$ is a function of class C^1. In addition, we now assume that

The range $R(L)$ of L is *closed* ; $\qquad (L_1)$

F is *convex* with $\nabla F : \mathbb{R}^n \longrightarrow \mathbb{R}$ $1-1$ and *onto* (for example, F strictly convex and such that $F(u)/|u| \to \infty$ as $|u| \to \infty$, cf. Remark 1.1). $\qquad (F_1)$

In this case, since L is selfadjoint, assumption (L_1) implies that we can decompose the space X as

$$X = N(L) \oplus R(L) \ ,$$

where $N(L)$ is the null space of L, $R(L)$ is the range of L, and the restriction $L|R(L) : D(L) \cap R(L) \longrightarrow R(L)$ has a bounded inverse $K = L^{-1} : R(L) \longrightarrow R(L)$. Moreover, in view of Proposition 1.1, assumption (F_1) implies that the

Legendre–Fenchel transform of F, $G = F^* : \mathbb{R}^n \longrightarrow \mathbb{R}$, is well defined and $\nabla G = (\nabla F)^{-1}$.

Now, by making the change of variable $v = \nabla F(u)$ in (P), that is, $u = \nabla G(v)$, we obtain the equivalent problem

$$L(\nabla G(v)) + v = 0 \quad , \quad v \in R(L),$$

or, by applying $K = L^{-1}$,

$$Kv + \nabla G(v) \in N(L) \quad , \quad v \in R(L). \tag{P^*}$$

The above problem (P^*) is *dual* to problem (P) and, similarly to problem (P), it has a *variational formulation*:

Find the critical points of the functional $\psi = \varphi^*$ given by

$$\psi(v) = \frac{1}{2}(Kv, v)_X + \int_\Omega G(v)\, dx \quad , \quad v \in R(L) \tag{ϕ^*}$$

(which we call the *dual action of Clarke–Ekeland*). In fact, in order to show that ψ is well defined, we will use the following proposition, where $F : \mathbb{R}^n \longrightarrow \mathbb{R}$ is assumed to satisfy a stronger condition than (F_1).

Proposition 3.1. *Assume conditions* (L_1) *and* (F_2), *where*

> $F \in C^1(\mathbb{R}^n, \mathbb{R})$ *is* strictly convex *and there exist constants* $0 < \alpha \leq \beta$, $\delta \in \mathbb{R}$, *such that*
>
> $$\alpha \frac{|u|^2}{2} - \delta \leq F(u) \leq \beta \frac{|u|^2}{2} + \delta \quad \forall u \in \mathbb{R}^n\ . \tag{F_2}$$

Then, the functional $\psi : R(L) \longrightarrow \mathbb{R}$ *given by* (ϕ^*) *is well defined and of class* C^1 *with*

$$\psi'(v) \cdot k = (Kv + \nabla G(v), k)_X \quad \forall v, k \in R(L).$$

In particular, $v \in R(L)$ *is a solution of* (P^*) *if and only if* v *is a critical point of* ψ.

Proof: In view of (F_2) and Remark 1.2, the dual action of Clarke–Ekeland, $G = F^*$, is well defined and satisfies

$$\frac{1}{\beta}\frac{|v|^2}{2} - \delta \leq G(v) \leq \frac{1}{\alpha}\frac{|v|^2}{2} + \delta \quad \forall v \in \mathbb{R}^n\ . \tag{3.1}$$

In particular, the second inequality above shows that the dual action ψ given by formula (ϕ^*) is well defined. Also, we have that $G \in C^1(\mathbb{R}^n, \mathbb{R})$ and G is *strictly convex* (i.e., ∇G is *strictly monotone*), since $F \in C^1(\mathbb{R}^n, \mathbb{R})$ is strictly convex, and $\nabla G = (\nabla F)^{-1}$. Therefore, since G satisfies (3.1), we obtain from Remark 1.1 that G^* is well defined. Moreover, given $u \in \mathbb{R}^n$, we have

$$G^*(u) = \sup_z [(u, z) - G(z)] = (u, (\nabla G)^{-1}(u)) - G((\nabla G)^{-1}(u)) ,$$

or

$$G^*(u) = (u, v) - G(v) , \tag{3.2}$$

where $v = (\nabla G)^{-1}(u)$, i.e., $u = \nabla G(v) = (\nabla F)^{-1}(v)$. Comparing (3.2) with (1.3) we conclude that $G^*(u) = F(u)$. Hence, using the first inequality in (F_2) with $u = \nabla G(v)$, it follows that

$$\alpha \frac{|u|^2}{2} - \delta \leq F(u) = G^*(u) = (u, v) - G(v) ,$$

so that (since $G(v) \geq -\delta$ from (3.1))

$$\alpha \frac{|\nabla G(v)|^2}{2} - \delta \leq (\nabla G(v), v) + \delta ,$$

which shows that $|\nabla G(v)|$ has linear growth in $|v|$:

$$|\nabla G(v)| \leq c|v| + d . \tag{3.3}$$

Finally, in view of (3.3), it is easy to check that the dual action ψ given in (ϕ^*) is of class C^1 on the closed subspace $R(L) \subset X = L^2(\Omega; \mathbb{R}^n)$ and

$$\begin{aligned} \psi'(v) \cdot k &= \int_\Omega [(Kv, k) + (\nabla G(v), k)] \, dx \\ &= (Kv + \nabla G(v), k)_X \quad \forall v, k \in R(L) . \end{aligned}$$

In particular, $v \in R(L)$ is such that $\psi'(v) = 0 \in R(L)^*$ if and only if $Kv + \nabla G(v) \in R(L)^\perp = N(L)$, that is, if and only if v is a solution of (P^*).

Remark 3.1. Note that, for $v \in R(L)$, the gradient $\nabla\psi(v)$ is given by $\nabla\psi(v) = Kv + Q\nabla G(v)$, where $Q : X \longrightarrow R(L)$ is the orthogonal projection of X onto $R(L)$.

Remark 3.2. As we have seen above, if we assume (L_1) and (F_2), then problem (P), whose variational formulation (cf. Proposition 2.1) is given by

$$\nabla\varphi(u) = 0 \ , \quad u \in D(L), \tag{VF}$$

is equivalent to the *dual problem* (P^*), whose variational formulation (cf. Proposition 3.1) is given by

$$\nabla\psi(v) = 0 \ , \quad v \in R(L) \tag{VF^*}$$

(where we recall that the *dual action* $\psi = \varphi^*$ is defined in (ϕ^*)). More precisely, we have seen that if $u \in D(L)$ is a critical point of φ (i.e., $u \in D(L)$ is a solution of (P)), then $v = -Lu \in R(L)$ is a critical point of ψ (i.e., $v \in R(L)$ is a solution of (P^*)). And, conversely, if $v \in R(L)$ is a critical point of ψ, then $u = \nabla G(v) \in D(L)$ is a critical point of φ (indeed, from $Kv + \nabla G(v) = w \in N(L)$ we obtain $L(Kv + \nabla G(v)) = 0$, that is, $v + Lu = 0$ with $u = \nabla G(v) = (\nabla F)^{-1}(v)$).

We point out that working with the dual action ψ instead of φ can be useful in situations in which the functional φ is *indefinite* (because the quadratic term $\frac{1}{2}(Lu, u)_X$ neither dominates nor is dominated by the term $\int_\Omega F(u)dx$), whereas ψ becomes a *coercive functional* (because the term $\int_\Omega G(v)dx$ dominates the quadratic term $\frac{1}{2}(Kv, v)_X$). We shall see examples of such situations in the following section.

4 Applications

In this section we first present a result due to Coron [28] on the existence of a nontrivial solution for the problem

$$Lu + \nabla F(u) = 0 \ , \quad u \in D(L) \subset X \ , \tag{P}$$

where, as before, $X = L^2(\Omega; \mathbb{R}^n)$ and $L : D(L) \subset X \longrightarrow X$ is a selfadjoint operator. Now, we assume that the essential spectrum of L is *empty*, that is, $\sigma(L) = \{\lambda_j \mid j \in \mathbb{Z}$ *(or* $\mathbb{N})\}$, where each λ_j is an isolated eigenvalue of finite multiplicity, and that $F \in C^1(\mathbb{R}^n, \mathbb{R})$ is *strictly convex* and satisfies $F(0) = \nabla F(0) = 0$. Clearly, in this case $u = 0$ is a solution of (P) and, in fact, is the *unique* solution if L is a *monotone operator*. Therefore, we will assume that $\sigma(L) \cap (-\infty, 0) \neq \emptyset$ and denote the first negative eigenvalue of L by λ_{-1}.

Theorem 4.1. *([28]) In addition to the above conditions, if F and L satisfy*

$$\alpha\frac{|u|^2}{2} - \delta \le F(u) \le \beta\frac{|u|^2}{2} + \delta \quad \forall u \in \mathbb{R}^n, \ \textit{with } 0 < \alpha \le \beta < -\lambda_{-1}, \ \delta \in \mathbb{R}, \tag{F_3}$$

$$\liminf_{u\to 0} \frac{2F(u)}{|u|^2} > -\lambda_{-1} \,, \tag{F_4}$$

$N(L-\lambda_{-1}) \subset L^\infty(\Omega;\mathbb{R}^n)$, *where* $N(L-\lambda_{-1})$ *is the* λ_{-1}*-eigenspace of* L , (L_2)

then problem (P) *has a nontrivial solution.*

Here we will present a proof that uses the dual action ψ of Clarke and Ekeland and which shows that one such nontrivial solution can be found by minimization of ψ over $R(L)$, cf. [30].

Lemma 4.2. *If* $L : D(L) \subset X \longrightarrow X$ *is a selfadjoint operator on* $X = L^2(\Omega;\mathbb{R}^n)$ *having an empty essential spectrum and* $F \in C^1(\mathbb{R}^n,\mathbb{R})$ *is a strictly convex function satisfying* (F_3), *then the dual action* $\psi : R(L) \longrightarrow \mathbb{R}$ *is* coercive.

Proof: In view of the assumption on L, the range $R(L)$ is closed and the restriction $L|D(L)\cap R(L) \longrightarrow R(L)$ possesses a *compact* inverse $K : R(L) \longrightarrow R(L)$. Moreover, through an eigenfunction expansion argument, K satisfies the estimate from below

$$(Kv,v)_X \geq \frac{1}{\lambda_{-1}}||v||_X^2 \quad \forall v \in R(L). \tag{4.1}$$

On the other hand, (F_3) implies that $G = F^*$ is well defined, satisfies (cf. (3.1))

$$\frac{1}{\beta}\frac{|v|^2}{2} - \delta \leq G(v) \leq \frac{1}{\alpha}\frac{|v|^2}{2} + \delta \quad \forall v \in R(L) \,,$$

and, in view of Proposition 3.1, the dual action $\psi : R(L) \longrightarrow \mathbb{R}$, given by

$$\psi(v) = \frac{1}{2}(Kv,v)_X + \int_\Omega G(v)\, dx$$

is well defined and of class C^1. Now, from (4.1) and (3.1) (recalled above), we obtain

$$\psi(v) \geq \frac{1}{2}\left(\frac{1}{\lambda_{-1}} + \frac{1}{\beta}\right)||v||_X^2 - \delta|\Omega| \quad \forall v \in R(L) \,.$$

Therefore, since $0 < \beta < -\lambda_{-1}$, the dual action ψ is coercive. □

Lemma 4.3. *Assume that the selfadjoint operator* L *is as in the previous lemma and satisfies the condition* (L_2). *In addition, assume that* $F \in C^1(\mathbb{R}^n,\mathbb{R})$ *is strictly convex,* $F(0) = \nabla F(0) = 0$, *and the conditions* (F_3), (F_4) *are satisfied.*[1] *Then, there exists* $\rho > 0$ *such that*

[1] For this lemma we do *not* need that β satisfy $\beta < -\lambda_{-1}$ in (F_3).

$$\psi(v) < 0 \quad \textit{for } v \in \Sigma_\rho := \{v \in N(L - \lambda_{-1}) \mid ||v||_X = \rho\} \ .$$

Proof: Condition (F_4) implies the existence of $\epsilon > 0$ and $\lambda^* > -\lambda_{-1}$ such that $F(u) \geq \lambda^*|u|^2/2$ for $|u| \leq \epsilon$. On the other hand, there exists $\rho' > 0$ such that $|\nabla G(v)| \leq \epsilon$ for $|v| \leq \rho'$. Therefore, since $G(v) = (u, v) - F(u)$ with $u = \nabla G(v)$, we obtain, for $|v| \leq \rho'$, that

$$\begin{aligned} G(v) &\leq \sup_{|u|\leq\epsilon} [(u,v) - \frac{\lambda^*}{2}|u|^2] \\ &\leq \sup_u [(u,v) - \frac{\lambda^*}{2}|u|^2] = \frac{1}{\lambda^*}\frac{|v|^2}{2} \ . \end{aligned}$$

Now, since we clearly have that

$$(Kv, v)_X = \frac{1}{\lambda_{-1}}||v||_X^2$$

for all $v \in N(L - \lambda_{-1})$, it follows from the previous estimate that

$$\psi(v) \leq \frac{1}{2}(\frac{1}{\lambda_{-1}} + \frac{1}{\lambda_*})||v||_X^2 < 0$$

for all $v \in N(L - \lambda_{-1})$ with $0 < ||v||_{L^\infty} \leq \rho'$. The proof is complete in view of the assumption (L_2) and the fact that $N(L - \lambda_{-1})$ is finite dimensional.□

Proof of Theorem 4.1: In view of Proposition 3.1 and Remark 3.2, it suffices to find a critical point $v_0 \neq 0$ of the dual action $\psi \in C^1(R(L), \mathbb{R})$. Since ψ is coercive by Lemma 4.2 and weakly l.s.c. (cf. Examples A and C in Chapter 2), it follows that $\psi(v_0) = \inf_{R(L)} \psi$ for some $v_0 \in R(L)$. Therefore, v_0 is a critical point of ψ and, since $\psi(v_0) < 0$ by Lemma 4.3, we have that $v_0 \neq 0$.

□

Next, we apply Theorem 4.1 to the problem of existence of a T-periodic solution for the classical second-order Hamiltonian system

$$\ddot{x} + \nabla V(x) = 0 \ , \tag{HS}$$

where $x(t) \in \mathbb{R}^n$ and $V \in C^1(\mathbb{R}^n, \mathbb{R})$ is a strictly convex function satisfying $V(0) = \nabla V(0) = 0$ and

$$\limsup_{|x|\to\infty} \frac{2V(x)}{|x|^2} < \frac{4\pi^2}{T^2} < \liminf_{|x|\to 0} \frac{2V(x)}{|x|^2} \ . \tag{V_1}$$

(Compare with Theorem 6.3.4.) In this case one can show that, in fact, the above system possesses a periodic solution with *minimal period* T.

Theorem 4.4. *([26]) Under the above conditions, the* (HS) *has a periodic solution with* minimal period T, *for each* $T > 0$.

Proof: Let $X = L^2(0,T;\mathbb{R}^n)$ and $L : D(L) \subset X \longrightarrow X$ be defined by

$$Lx = \ddot{x} \quad \forall x \subset D(L) \ ,$$

$D(L) = \{x \in X \mid x,\ \dot{x} \text{ abs. continuous},\ x(0) = x(T),\ \dot{x}(0) = \dot{x}(T) \text{ and } \ddot{x} \in X\}$. Then L is selfadjoint, $\sigma(L) = \{-\frac{4\pi^2 j^2}{T^2} \mid j = 0, 1, 2, \dots\}$ and each $\lambda_{-j} = -4\pi^2 j^2/T^2$ is an eigenvalue of L having finite multiplicity. The eigenfunctions corresponding to λ_{-j} are given by

$$(\cos\frac{2\pi jt}{T})e \ , \quad (\sin\frac{2\pi jt}{T})e \ ,$$

where $0 \neq e \in \mathbb{R}^n$, so that condition (L_2) is satisfied. Also, assumption (V_1) implies (F_3) and (F_4). Therefore, in view of Theorem 4.1, the Hamiltonian system (HS) possesses a nontrivial T-periodic solution x_0 such that $v_0 = -Lx_0 \neq 0$ minimizes the dual action ψ over $R(L)$.

It remains to show that v_0 has minimal period T. Assume, by negation, that v_0 has period $\frac{T}{m}$, where $m \geq 2$. If we define $\hat{v}_0(t) = v_0(t/m)$, we readily obtain that $\hat{v}_0 \in R(L)$ (i.e., $\int_0^T \hat{v}_0 dt = 0$) and

$$\begin{aligned}\psi(\hat{v}_0) &= \frac{m^2}{2}(Kv_0, v_0)_X + \int_0^T G(v_0)\, dt \\ &= \psi(v_0) + \frac{m^2-1}{2}(Kv_0, v_0)_X \ ,\end{aligned}$$

so that $\psi(\hat{v}_0) < \psi(v_0)$, since $m \geq 2$ and K is negative definite on $R(L)$. This contradicts the fact that v_0 minimizes ψ. Therefore, v_0 indeed has minimal period T. □

8

Critical Points under Symmetries

1 Introduction

As we have mentioned briefly in Section 4.1, the central idea behind the the so-called Lusternik–Schnirelman theory is to obtain critical values of a given functional $\varphi \in C^1(X, \mathbb{R})$ as minimax (or maximin) values of φ over suitable classes $\mathcal{A}_k$ of subsets of X,

$$c_k = \inf_{A \in \mathcal{A}_k} \sup_{x \in A} \varphi(x) \ , \quad k = 1, 2, \ldots \ ,$$

where, originally, the classes $\mathcal{A}_k$ were based on the topological notion of *category of Lusternik–Schnirelman.*

When the problem presents symmetries, that is, when there exists some group G *acting* in a continuous manner on the space X and the functional φ is *invariant* under that action, then it usually happens that φ possesses *many* critical points. The typical example of such a multiplicity situation is a classical theorem of Lusternik, which guarantees the existence of at least $(n+1)$ distinct pairs of critical points for any functional $\varphi \in C^1(S^n, \mathbb{R})$ that is *even* ([53]). Moreover, when in presence of symmetries, it is possible to choose the classes $\mathcal{A}_k$ by using topological notions which are more manageable than the category.

In this section we will be considering this question of *multiplicity versus symmetry.* Our presentation will follow the work [55] of Mawhin–Willem, and we will start by introducing the notion of a *G-index theory.* The most well-known examples of such theories refer to the cases where $G = \mathbb{Z}_2$ or $G = S^1$. In the first case, we have the notion of *genus* due to Krasnoselskii [47] (also see Coffman [27]). In the second case, we have the *cohomological index* of Fadell–Rabinowitz [41] (see also Fadell–Husseini–Rabinowitz [42]) and the *geometric index* of Benci [10].

2 The Lusternik–Schnirelman Theory

Let X be a Banach space and let G be a compact topological group. Assume that $\{T(g) \mid g \in G\}$ is an *isometric representation* of G on X, that is, for each $g \in G$, $T(g) : X \longrightarrow X$ is an isometry and the following properties hold:

(i) $T(g_1 + g_2) = T(g_1) \circ T(g_2) \quad \forall g_1, g_2 \in G$,
(ii) $T(0) = I$ (the identity on X) ,
(iii) $(g, u) \mapsto T(g)u$ is continuous.

The *orbit* of an element $u \in X$ is the set $\mathcal{O}(u) = \{T(g)u \mid g \in G\}$, and a subset $A \subset X$ is said to be *invariant* if $T(g)A = A$ for all $g \in G$, that is, if A is a union of orbits. On the other hand, a functional $\varphi : X \longrightarrow \mathbb{R}$ is called *invariant* if φ is constant on each orbit of X, that is, $\varphi \circ T(g) = \varphi$ for all $g \in G$. And, given invariant subsets A_1, A_2, a mapping $R : A_1 \longrightarrow A_2$ is said to be *equivariant* if $R \circ T(g) = T(g) \circ R$ for all $g \in G$ (i.e., R maps the *orbit of* u pointwise onto the *orbit of* $R(u)$).

Let us denote by $\mathcal{A}$ the class of all *closed* and *invariant* subsets of X, $\mathcal{A} = \{A \subset X \mid A \text{ is closed and } T(g)A = A \ \forall g \in G \}$.

Definition. A G-index on X (w.r.t. $\mathcal{A}$) is a mapping ind $: \mathcal{A} \longrightarrow \mathbb{N} \cup \{\infty\}$ satisfying the following properties (for any $A, N, A_1, A_2 \in \mathcal{A}$):

(a) ind $(A) = 0$ if and only if $A = \emptyset$;
(b) If $R : A_1 \longrightarrow A_2$ is continuous and equivariant, then ind $(A_1) \leq$ ind (A_2);
(c) ind $(A_1 \cup A_2) \leq$ ind $(A_1) +$ ind (A_2);
(d) If $A \in \mathcal{A}$ is compact, then there exists a neighborhood N of A such that $N \in \mathcal{A}$ and ind $(N) =$ ind (A).

Proposition 2.1. *Let X be a Banach space, $G = \mathbb{Z}_2 = \{0, 1\}$, and define $T(0) = I$, $T(1) = -I$ (where $I : X \longrightarrow X$ is the identity map on X). Given any closed, symmetric (w.r.t. the origin) subset $A \in \mathcal{A}$, define $\gamma(A) = k \in \mathbb{N}$ if k is the smallest integer such that there exists some* odd, continuous mapping *$\Phi : A \longrightarrow \mathbb{R}^k \backslash \{0\}$. Also define $\gamma(A) = \infty$ if no such mapping exists, and set $\gamma(\emptyset) = 0$. Then, the mapping $\gamma : \mathcal{A} \longrightarrow \mathbb{N} \cup \{\infty\}$ is a $\mathbb{Z}_2$-index on X.*

Remark. This $\mathbb{Z}_2$-index is called the *genus* and it was originally introduced by Krasnoselskii [47]. The definition given above is due to Coffman [27] (and it is equivalent to the original definition of Krasnoselskii).

Proof of Proposition 2.1: First, observe that $A \subset X$ is $\mathbb{Z}_2$-invariant if and only if $u \in A$ implies $-u \in A$, that is, A is symmetric w.r.t. the origin in X.

Thus $\mathcal{A} = \{A \subset X \mid A \text{ is closed and } -A = A\}$. On the other hand, given invariant subsets A_1, A_2, a mapping $R : A_1 \longrightarrow A_2$ is equivariant if and only if R is continuous and *odd*. Let us check properties (a) – (d) in the definition of a G-index.

(a) This property is immediate from the definition of γ.

(b) Since there is nothing to prove if $\gamma(A_2) = \infty$, let us assume that $\gamma(A_2) = k < \infty$, so that there exists a continuous, odd mapping $\Phi : A_2 \longrightarrow \mathbb{R}^k \backslash \{0\}$. Then, since the mapping $\Phi \circ R : A_1 \longrightarrow \mathbb{R}^k \backslash \{0\}$ is also continuous and odd, it follows from the definition that $\gamma(A_1) \leq k$.

(c) Again, there is nothing to prove if $\gamma(A_1) = \infty$ or $\gamma(A_2) = \infty$. Therefore, we assume that $\gamma(A_1) = k_1 < \infty$ and $\gamma(A_2) = k_2 < \infty$, so that there exist continuous, odd mappings $\Phi_j : A_j \longrightarrow \mathbb{R}^{k_j} \backslash \{0\}$, $j = 1, 2$. Let $\Psi_j : X \longrightarrow \mathbb{R}^{k_j}$ be the continuous, odd extension of Φ_j, obtained by taking the *odd part* $\Psi_j = \frac{1}{2}[\hat{\Phi}_j(u) - \hat{\Phi}_j(-u)]$ of some extension $\hat{\Phi}_j$ of Φ_j (guaranteed by Tietze extension theorem). Then, the mapping $\Psi : X \longrightarrow \mathbb{R}^{k_1+k_2}$ defined by

$$\Psi(u) = (\Psi_1(u), \Psi_2(u))$$

is such that $\Psi(A_1 \cup A_2) \subset \mathbb{R}^{k_1+k_2} \backslash \{0\}$ because $\Psi_j(A_j) \subset \mathbb{R}^{k_j} \backslash \{0\}$. It follows that the restriction (of Ψ to $A_1 \cup A_2$)

$$\Psi|(A_1 \cup A_2) : A_1 \cup A_2 \longrightarrow \mathbb{R}^{k_1+k_2} \backslash \{0\}$$

is a continuous, odd mapping and, hence, $\gamma(A_1 \cup A_2) \leq k_1 + k_2$.

(d) As before, it suffices to consider the case $\gamma(A) = k < \infty$ so that there exists an odd mapping $\Phi \in C(A, \mathbb{R}^k \backslash \{0\})$. Let $\Psi \in C(X, \mathbb{R}^k)$ be an odd extension of Φ, obtained as in (c). Since $0 \notin \Psi(A) = \Phi(A)$ and A is compact by assumption, there exists $\delta > 0$ such that $0 \notin \Psi(A_\delta)$, where $A_\delta := \{u \in X \mid \operatorname{dist}(u, A) \leq \delta\}$. Then, the set $N = A_\delta$ is a closed neighborhood of A, which is clearly invariant, and $\gamma(N) \leq k$ by construction. On the other hand, property *(b)* implies that $k = \gamma(A) \leq \gamma(N)$ since the inclusion $A \subset N$ is obviously continuous and odd. Thus, $\gamma(N) = \gamma(A) = k$. □

Our next proposition provides several examples of the genus of subsets of X.

Proposition 2.2. *(i) If $C \subset X$ is closed and $C \cap (-C) = \emptyset$, then $\gamma(C \cup (-C)) = 1$;*

(ii) If $A \in \mathcal{A}$ and there exists an odd homeomorphism $h : A \longrightarrow S^{k-1}$, then $\gamma(A) = k$ (in particular, $\gamma(S^{k-1}) = k$);

(iii) If $A \in \mathcal{A}$ is such that $0 \notin A$ and $\gamma(A) \geq 2$, then A has infinitely many points.

Proof: (*i*) Simply take $\Phi : C \cup (-C) \longrightarrow \mathbb{R}\backslash\{0\}$ defined by $\Phi(u) = 1$ if $u \in C$, $\Phi(u) = -1$ if $u \in -C$.

(*ii*) Since $h : A \longrightarrow S^{k-1}$ is continuous and odd, we have that $\gamma(A) \leq k$. On the other hand, if $\gamma(A) = j < k$, then there exists an odd mapping $\Phi \in C(A, \mathbb{R}^j\backslash\{0\})$, hence an odd mapping $\Psi := \Phi \circ h^{-1} \in C(S^{k-1}, \mathbb{R}^j\backslash\{0\})$. Since $j < k$, the mapping Ψ (considered as a mapping taking values into $\mathbb{R}^k$) is clearly homotopic to a constant mapping $\Psi_0 = u_0$, where $0 \neq u_0 \in \mathbb{R}^k\backslash\mathbb{R}^j$, say $|u_0| > 1$.[1] It follows from properties (*v*), (*ii*) in Section 5.1 that, necessarily we have the degree $\deg(\Psi, B_k, 0) = 0$, where B_k is the unit ball in $\mathbb{R}^k$. But this contradicts the Borsuk theorem (cf. [57]): "If $\Omega \subset \mathbb{R}^k$ is a bounded, open and symmetric set with $0 \in \Omega$ and if $\Psi \in C(\partial\Omega, \mathbb{R}^k\backslash\{0\})$ is an odd mapping, then $\deg(\Psi, \Omega, 0)$ is an *odd integer*."

(*iii*) It suffices to observe that if $A \in \mathcal{A}$ is a set with a finite number of points and $0 \notin A$, then we can write A as a union $A = C \cup (-C)$ with $C \cap (-C) = \emptyset$, so that $\gamma(A) = 1$ in view of (*i*). □

Remark 2.1. An example of an index theory in the case $G = S^1$ will be presented in Chapter 9.

3 The Basic Abstract Multiplicity Result

In this section we will state and prove the basic multiplicity result in the Lusternik–Schnirelman theory in the presence of symmetries. For that, we will need a deformation theorem which *preserves* the symmetries, that is, a deformation $\eta \in C([0,1] \times X, X)$ as in Chapter 3 satisfying the additional condition that $\eta(t, \cdot) : X \longrightarrow X$ is equivariant for each $t \in [0,1]$. On the other hand, the existence theorem of an *equivariant deformation* will be based on the notion of an *equivariant pseudo-gradient* for an invariant functional $\varphi \in C^1(X, \mathbb{R})$, whose existence will be established after two preliminary lemmas (cf. [55]).

Lemma 3.1. *Let $\varphi \in C^1(X, \mathbb{R})$ be an invariant functional. Then, for any $g \in G$ and $u, h \in X$, one has:*

[1] We are identifying the points $x \in \mathbb{R}^j$ with $(x, 0) \in \mathbb{R}^k$, where 0 is the origin in $\mathbb{R}^{k-j}$.

(*i*) $\varphi'(T(g)u) \cdot h = \varphi'(u) \cdot T(-g)h$;
(*ii*) $||\varphi'(T(g)u)|| = ||\varphi'(u)||$.

Proof: Since each $T(g) : X \longrightarrow X$ is a surjective isometry (i.e., a unitary mapping) with inverse $T(-g)$, (*ii*) follows readily from (*i*). So, it suffices to check (*i*). Indeed, the invariance of φ and the definition of $\varphi'(v) \cdot k$ imply

$$\begin{aligned}\varphi'(T(g)u) \cdot h &= \lim_{t\to 0} \frac{1}{t}[\varphi(T(g)(u + tT(-g)h)) - \varphi(T(g)u)] \\ &= \lim_{t\to 0} \frac{1}{t}[\varphi(u + tT(-g)h) - \varphi(u)] ,\end{aligned}$$

which proves (*i*). □

Lemma 3.2. *Let* $\omega : Y \longrightarrow X$ *be a locally Lipschitzian mapping, where* X *and* Y *are metric spaces. Given a compact subset* $A \subset Y$ *there exists* $\delta > 0$ *such that* ω *is (globally) Lipschitzian on* $A_\delta = \{u \in Y \mid dist(u, A) \leq \delta\}$.

Proof: We leave it as an exercise for the reader (or see [55]). □

Proposition 3.3. *An* invariant *functional* $\varphi \in C^1(X, \mathbb{R})$ *possesses an* equivariant *pseudo-gradient, that is, a locally Lipschitzian mapping* $v : Y \longrightarrow X$, *where* $Y = \{u \in X \mid \varphi'(u) \neq 0\}$, *satisfying:*

(*i*) $||v(u)|| \leq 2||\varphi'(u)||$;
(*ii*) $\varphi'(u) \cdot v(u) \geq ||\varphi'(u)||^2$;
(*iii*) v *is* equivariant.

Proof: We show the existence of an equivariant pseudo-gradient (for φ) starting from a corresponding pseudo-gradient ω, whose existence shall be assumed without a proof, as in Section 3.1. Indeed, define

$$v(u) = \int_G T(-g)\omega(T(g)u)\, dg \ , \quad u \in Y \ ,$$ [2]

where dg is the Haar measure on G, normalized so that $\int_G dg = 1$ (see [48]). We have:

$$||v(u)|| \leq \int_G ||\omega(T(g)u)||\, dg \leq \int_G 2||\varphi'(T(g)u)||\, dg = 2||\varphi'(u)|| \ ,$$

[2] In the case $G = \mathbb{Z}_2$ with $T(0) = I$, $T(1) = -I$, we observe that $v(u)$ is simply the *odd part* of $\omega(u)$, i.e., $v(u) = \frac{1}{2}[\omega(u) - \omega(-u)]$.

where we used Lemma 3.1 *(ii)* in the last equality. This proves item *(i)*.

Next, we observe that

$$\begin{aligned}\varphi'(u)\cdot v(u) &= \int_G \varphi'(u)\cdot[T(-g)\omega(T(g)u)]\,dg\\ &= \int_G \varphi'(T(g)u)\cdot\omega(T(g)u)\,dg\\ &\geq \int_G ||\varphi'(T(g)u)||^2\,dg = ||\varphi'(u)||^2\,,\end{aligned}$$

where we used Lemma 3.1 *(i)* and *(ii)*, respectively, in the second and last equality. This proves item *(ii)*.

The proof that $v(u)$ is an equivariant pseudo-gradient follows from

$$\begin{aligned}v(T(\hat{g})u)) &= \int_G T(-g)\omega(T(g+\hat{g})u)\,dg\\ &= \int_G T(\hat{g})T(-g-\hat{g})\omega(T(g+\hat{g})u)\,dg = T(\hat{g})v(u)\,,\end{aligned}$$

where we used the translation invariance of the Haar measure in the last equality above.

Finally, we show that v is locally Lipschitzian. For fixed $u\in Y$, let $A=\{T(g)u\mid g\in G\}$ be the orbit of u and $\delta>0$ be given by Lemma 3.2, so that ω is Lipschitzian on A_δ, with (say) constant l. Then, for arbitrary $u_1,u_2\in A_\delta(u)$, we have

$$\begin{aligned}||v(u_1)-v(u_2)|| &\leq \int_G ||\omega(T(g)u_1)-\omega(T(g)u_2)||\,dg\\ &\leq \int_G l||T(g)u_1-T(g)u_2||\,dg = l||u_1-u_2||\,. \qquad \square\end{aligned}$$

Theorem 3.4. (Theorem of the Equivariant Deformation) *Let $\varphi\in C^1(X,\mathbb{R})$ be an invariant functional satisfying the Palais–Smale condition (PS). If U is an invariant open neighborhood of K_c, $c\in\mathbb{R}$, then, for all ϵ sufficiently small, there exists $\eta\in C([0,1]\times X,X)$ such that (for any $u\in X$ and $t\in[0,1]$):*

(i) $\eta(0,u)=u$,
(ii) $\eta(t,u)=u$ *if* $u\notin\varphi^{-1}[c-2\epsilon,c+2\epsilon]$,
(iii) $\eta(1,\varphi^{c+\epsilon}\setminus U)\subset\varphi^{c-\epsilon}$,
(iv) $\eta(t,\cdot):X\longrightarrow X$ *is an equivariant homeomorphism.*

Proof: The only difference between this result and Theorem 3.2.3 is that we must show that the deformation $\eta(t,\cdot)$ can be chosen to be equivariant in this case in which both the functional φ and the neighborhood $U \supset K_c$ are invariant.

Indeed, since φ is invariant, we may assume by the previous proposition that φ has an *equivariant* pseudo-gradient v. Moreover, the invariance of the neighborhood $U \supset K_c$ allows us to choose the cut-off function $\rho : X \longrightarrow \mathbb{R}$ (used in the proof of Theorem 3.2.1) to be invariant. It suffices to take

$$\rho(u) = \frac{\mathrm{dist}(u, X\backslash A)}{\mathrm{dist}(u, X\backslash A) + \mathrm{dist}(u, B)}$$

and recall that each $T(g)$ is an isometry. Then, the vector field $f(u) = -\rho(u)v(u)/||v(u)||$ is equivariant and, in view of the uniqueness of solution for the Cauchy problem $\frac{dw}{dt} = f(w)$, $w(0) = u$, we conclude that $w(t, \cdot)$ and $\eta(t, \cdot) = w(\delta t, \cdot)$ are equivariant mappings (indeed, if $w(t, u)$ is the solution of the above Cauchy problem then $w_g(t) := T(g)w(t, u)$ is the solution of the Cauchy problem with initial condition $w_g(0) = T(g)u$, that is, $T(g)w(t, u) = w(t, T(g)u)$). □

Now we will assume that, besides an isometric representation $\{T(g)\}$ of G in X, we also have a G-index ind $: \mathcal{A} \longrightarrow \mathbb{N} \cup \{\infty\}$. Then, we can "classify" the *compact, invariant* subsets of X by defining, for each $j = 1, 2, \dots$, the class

$$\mathcal{A}_j = \{A \subset X \mid A \text{ is compact, invariant and } \mathrm{ind}\,(A) \geq j\ \} \,.$$

And, for a given functional $\varphi : X \longrightarrow \mathbb{R}$, we define $c_j = c_j(\varphi)$, $j = 1, 2, \dots$, by the formula

$$c_j = \inf_{A \in \mathcal{A}_j} \max_{u \in A} \varphi(u) \,. \tag{3.1}$$

(Note that we may have $c_j = -\infty$.) Since $\mathcal{A}_1 \supset \mathcal{A}_2 \supset \cdots$, we clearly have

$$-\infty \leq c_1 \leq c_2 \leq \cdots$$

Theorem 3.5. (Basic Abstract Multiplicity Theorem) *Let $\varphi \in C^1(X, \mathbb{R})$ be an invariant functional satisfying (PS). If $c_j > -\infty$ for some $j \geq 1$, then c_j is a critical value of φ. More generally, if $c_k = c_j = c > -\infty$ for some $k \geq j$, then ind$(K_c) \geq k - j + 1$.*

Proof: First, we observe that K_c is an invariant set, since φ is an invariant functional, and K_c is compact in view of the Palais–Smale condition. We now show that if $c_k = c_j = c > -\infty$ for some $k \geq j$, then ind$\,(K_c) \geq k - j + 1$.

Let $N \supset K_c$ be a closed, invariant neighborhood such that $\text{ind}\,(N) = \text{ind}\,(K_c)$, whose existence is implied by property (d) in the definition of G-index. Then the interior $U := \text{int}\,(N)$ of N is an open, invariant neighborhood of K_c, so that we may apply Theorem 3.4 to conclude the existence of $\epsilon > 0$ and $\eta \in C([0,1] \times X, X)$ satisfying properties (i)–(iv) of that theorem.

Now, by definition of $c = c_k$, take $A \in \mathcal{A}_k$ such that $\max_A \varphi \leq c + \epsilon$ and define the compact set $B = A \backslash U$. Then, properties (b), (c) in the definition of a G-index imply that

$$\begin{aligned} k &\leq \text{ind}\,(A) \\ &\leq \text{ind}\,(B) + \text{ind}\,(N) \\ &= \text{ind}\,(B) + \text{ind}\,(K_c)\ . \end{aligned} \tag{3.2}$$

And, since $B \subset \varphi^{c+\epsilon} \backslash U$, we obtain that $C = \eta(1, B) \subset \varphi^{c-\epsilon}$ by Theorem 3.4.

On the other hand, since $\eta(1, \cdot)$ is an equivariant mapping and B is a compact, invariant set, it follows that the set C is also compact and invariant. And, as we just saw, $\max_C \varphi \leq c - \epsilon$. Therefore, by definition of c_j, we necessarily have $\text{ind}\,(C) \leq j - 1$. Moreover, by property (b) in the definition of a G-index, we have

$$\text{ind}\,(B) \leq \text{ind}\,(C) \leq j - 1\ . \tag{3.3}$$

Finally, (3.2) and (3.3) imply that $\text{ind}\,(K_c) \geq k - j + 1$, as we wished to show. $\square$

One of the consequences of the above result is the following theorem due to Clark:

Theorem 3.6. *([25]) Let $\varphi \in C^1(X, \mathbb{R})$ be an* even *functional satisfying the Palais–Smale condition (PS). Suppose that*

(i) φ is bounded from below;
(ii) There exists a compact, symmetric set $K \in \mathcal{A}$ such that $\gamma(K) = k$ and

$$\sup_K \varphi < \varphi(0)\ .$$

Then, φ possesses at least k distinct pairs of critical points with corresponding critical values less than $\varphi(0)$.

Proof: Consider the c_j's defined in (3.1) where, in the present situation, $\mathcal{A}_j = \{A \subset X \mid A \text{ is compact, symmetric and } \gamma(A) \geq j\ \}$:

$$c_1 \leq c_2 \leq \cdots \leq c_k \leq \cdots\ .$$

Then, (i) implies that $c_1 = \inf_X \varphi > -\infty$ and (ii) implies that $c_k < \varphi(0)$.

Therefore, by Theorem 3.5, each c_j is a critical value of φ. If all the c_j's $(j = 1, \dots, k)$ are distinct we obtain at least k pairs of critical points with values less that $\varphi(0)$. On the other hand, if $c_i = c_j = c$ for some $1 \leq i < j \leq k$, then $\gamma(K_c) \geq j - i + 1 \geq 2$ and, since $0 \notin K_c$ (as $c \leq c_k < \varphi(0)$), Proposition 2.2 (iii) implies that K_c possesses infinitely many points in this case. □

4 Application to a Problem with a $\mathbb{Z}_2$-Symmetry

Let us again consider the Dirichlet problem (cf. Application A, Section 4.3)

$$\begin{cases} -\Delta u = u^3 & \text{in } \Omega \\ \quad u = 0 & \text{on } \partial\Omega \ , \end{cases} \tag{4.1}$$

where $\Omega \subset \mathbb{R}^3$ is bounded domain with smooth boundary. As we know, the functional

$$\varphi(u) = \int_\Omega [\frac{1}{2}|\nabla u|^2 - \frac{1}{4}u^4] \ dx \tag{4.2}$$

is well defined and of class C^1 on the Sobolev space $X = H_0^1$, and its critical points are precisely the solutions of (4.1). In Section 4.3 we found a nontrivial solution through the mountain-pass theorem. We now show that, in fact, (4.1) has an infinite number of solutions (cf. Ambrosetti–Rabinowitz [9]).

Theorem 4.1. *Problem (4.1) has infinitely many (classical) solutions.*

The original proof of Ambrosetti and Rabinowitz uses a *symmetric* version of the mountain-pass theorem. Here, we shall present another proof (cf. Castro [22]) which uses Clark's Theorem 3.6 applied to an associated functional ψ, which is *bounded from below* and whose set of critical points is related to the set of critical points of φ. For that, consider the functional

$$\psi(u) = \left(\int_\Omega |\nabla u|^2 \ dx\right)^3 - \int_\Omega u^4 \ dx = ||u||^6 - ||u||_{L^4}^4 \ , \tag{4.3}$$

which, similarly to φ, is well defined and of class C^1 on the space $X = H_0^1(\Omega)$, with derivative given by

$$\psi'(u) \cdot v = 6||u||^4 \int_\Omega \nabla u \cdot \nabla v \ dx - 4\int_\Omega u^3 v \ dx \quad , \quad \forall u, v \in X \ . \tag{4.4}$$

Then we have that

$$\nabla\psi(u) = 6||u||^4 u - T(u) \ , \tag{4.5}$$

where $T : X \longrightarrow X$ is the compact operator defined by $\langle T(u), v\rangle = 4\int_\Omega u^3 v dx$ $\forall u, v \in X$.

Lemma 4.2. *(i) ψ is* bounded from below*;*
(ii) ψ satisfies the Palais–Smale condition (PS).

Proof: (*i*) In view of the Sobolev embedding $X = H_0^1 \subset L^4$, we have

$$\psi(u) \geq ||u||^6 - c||u||^4 \ , \quad u \in X \ ,$$

so that ψ is bounded from below.

(*ii*) Let $(u_n) \subset X$ be such that $|\psi(u_n)| \leq C$, $\nabla\psi(u_n) \to 0$ in X. Then,

$$C \geq \psi(u_n) \geq ||u_n||^6 - c||u_n||^4 \ , \tag{4.6}$$

$$6||u_n||^4 u_n - T(u_n) \to 0 \ , \tag{4.7}$$

and from (4.6) we conclude that $||u_n||$ is bounded. So, we may assume (passing to a subsequence, if necessary) that there exist $\hat{u} \in X$ and $a \geq 0$ such that $u_n \rightharpoonup \hat{u}$ weakly in X and $||u_n|| \to a$. If $a = 0$ there is nothing to prove. If $a > 0$, then $||u_n|| > 0$ for all n sufficiently large and we can write

$$u_n = \frac{1}{6||u_n||^4}[6||u_n||^4 u_n - T(u_n) + T(u_n)] \ .$$

Therefore, we conclude from (4.7) that $u_n \to \frac{1}{6a^4}[0 + T(\hat{u})]$. □

Lemma 4.3. *If $0 \neq u \in X$ is a critical point of ψ, then $v = \frac{2u}{\sqrt{6}||u||^2}$ is a critical point of φ.*

Proof: A function $0 \neq u \in X$ is a critical point of ψ if and only if $u \neq 0$ is a *weak* solution of the problem $6||u||^4 \Delta u + 4u^3 = 0$ in Ω, $u = 0$ on $\partial\Omega$, that is, $\frac{6\sqrt{6}||u||^6}{2}[\Delta v + v^3] = 0$ in Ω, $v = 0$ on $\partial\Omega$, where $v = \frac{2u}{\sqrt{6}||u||^2}$. Therefore, $v = \frac{2u}{\sqrt{6}||u||^2}$ is a critical point of φ. □

Proof of Theorem 4.1: As we mentioned earlier, we will apply Clark's Theorem 3.6 to the functional ψ defined in (4.3). By Lemma 4.2 we know that ψ satisfies condition (PS) and

(*i*) ψ is bounded from below.

Next, given an arbitrary $k \in \mathbb{N}$, we will show that

(ii) there exists a compact, symmetric set $K \in \mathcal{A}$ such that $\gamma(K) = k$ and

$$\sup_K \psi < 0 \ .$$

In fact, let $X_k = \text{span}\,\{\phi_1, \dots, \phi_k\}$ be the subspace spanned by the first k eigenfunctions of the eigenvalue problem $-\Delta u = \lambda u$ in Ω, $u = 0$ on $\partial\Omega$. Then, since X_k is finite dimensional and $X_k \subset L^4$, there exists $a > 0$ such that

$$a||u||_{L^4} \leq ||u||_X \leq \frac{1}{a}||u||_{L^4} \quad \forall u \in X_k \ .$$

Therefore, we obtain

$$\psi(u) = ||u||_X^6 - ||u||_{L^4}^4 \leq A||u||_{L^4}^6 - ||u||_{L^4}^4 = ||u||_{L^4}^4(A||u||_{L^4}^2 - 1) \ ,$$

so, for $0 < ||u||_{L^4}^2 \leq \frac{1}{2A} = \delta^2$, it follows that

$$\psi(u) \leq -\frac{1}{2}||u||_{L^4}^4 < 0 \ .$$

Now, if we define the set $K = \{u \in X_k \mid \frac{a\delta}{2} \leq ||u||_X \leq a\delta\}$, then $K \in \mathcal{A}$ and $\sup_K \psi < 0$. Moreover, since X_k is isomorphic to $\mathbb{R}^k$, we may identify K with a *annulus* $\widetilde{K}$ in $\mathbb{R}^k$, and the inclusions $S^{k-1}(b) \subset \widetilde{K} \subset \mathbb{R}^k\backslash\{0\}$ (for some $b > 0$) show that $\gamma(K) = k$. Therefore, condition (ii) is satisfied and Clark's theorem implies the existence of at least k distinct pairs of critical points for the functional ψ. Since k is arbitrary, we obtain infinitely many critical points for ψ.

Finally, in view of Lemma 4.3, we conclude that the functional φ possesses, together with ψ, infinitely many critical points. And, as we know, each critical point of φ is a classical solution of problem (4.1). □

9

Problems with an S^1-Symmetry

1 A Geometric S^1-index

In this chapter we present an index theory for the case in which $G = S^1$. It is an extension (cf. [55]) of the geometric S^1-index due to Benci [10]. Also see Fadell–Rabinowitz [41].

Let X be a Banach space and suppose that $\{T(\theta) \mid \theta \in S^1\}$ is an isometric representation of S^1 in X. As before, we denote by $\mathcal{A}$ the class of all closed subsets of X which are invariant under the given representation, that is,

$$\mathcal{A} = \{A \subset X \mid A \text{ is closed and } T(\theta)A = A \ \forall \theta \in S^1 \ \}.$$

Proposition 1.1. *Given $A \in \mathcal{A}$, define* $\operatorname{ind}(A) = k$ *if k is the smallest integer for which there exists a mapping $\Phi \in C(A, \mathbb{C}^k \backslash \{0\})$ and an integer $n \in \mathbb{N} \backslash \{0\}$ satisfying the relation*

$$\Phi(T(\theta)u) = e^{in\theta}\Phi(u) \ \ \forall \theta \in S^1 \ , \quad \forall u \in X \ . \tag{1.1}$$

Moreover, define $\operatorname{ind}(A) = \infty$ *in case no such mapping exists, and define* $\operatorname{ind}(\emptyset) = 0$. *Then,* $\operatorname{ind} : \mathcal{A} \longrightarrow \mathbb{N} \cup \{\infty\}$ *is an S^1-index on X.*

Proof: We must verify properties (a)–(d) presented in Section 8.2:

(a) This property is immediate.

(b) We may assume $\operatorname{ind}(A_2) < \infty$, so that there exists a continuous mapping $\Phi : A_2 \longrightarrow \mathbb{C}^k \backslash \{0\}$ and some $n \in \mathbb{N} \backslash \{0\}$ satisfying the relation (1.1). From this and the fact that $R : A_1 \longrightarrow A_2$ is an equivariant mapping, we conclude that $\Psi = \Phi \circ R : A_1 \longrightarrow \mathbb{C}^k \backslash \{0\}$ is also continuous and satisfies (1.1). Thus, $\operatorname{ind}(A_1) \leq k = \operatorname{ind}(A_2)$.

(c) As before, we may assume that $\text{ind}\,(A_1) = k_1 < \infty$ and $\text{ind}\,(A_2) = k_2 < \infty$. Therefore, there exist continuous mappings $\Phi_j : A_j \longrightarrow \mathbb{C}^{k_j}\backslash\{0\}$ and $n_j \in \mathbb{N}\backslash\{0\}$, $j = 1, 2$, satisfying the relation (1.1). Let $\hat{\Phi}_j : X \longrightarrow \mathbb{C}^{k_j}$ be a continuous extension of Φ_j obtained by the Tietze extension theorem and define

$$\Psi_j(u) = \frac{1}{2\pi} \int_0^{2\pi} e^{-in_j\theta} \hat{\Phi}_j(T(\theta)u)\, d\theta \ , \quad j = 1, 2.$$

Then $\Psi_j : X \longrightarrow \mathbb{C}^{k_j}$ is a continuous extension of Φ_j satisfying (1.1) and, defining

$$\Psi(u) = (\Psi_1(u)^{n_2}, \Psi_2(u)^{n_1}) \ , \quad u \in X \ ,$$

we obtain $\Psi(A_1 \cup A_2) \subset \mathbb{C}^{k_1+k_2}\backslash\{0\}$ (since $\Psi_j(A_j) = \Phi_j(A_j) \subset \mathbb{C}^{k_j}\backslash\{0\}$) so that

$$\Psi(T(\theta)u) = ((e^{in_1\theta}\Psi_1(u))^{n_2}, (e^{in_2\theta}\Psi_2(u))^{n_1}) = e^{in_1n_2\theta}\Psi(u) \ .$$

Therefore, we have that $\Psi \in C(A_1 \cup A_2, \mathbb{C}^{k_1+k_2}\backslash\{0\})$ and Ψ satisfies (1.1) with $n = n_1n_2$, and it follows that $\text{ind}\,(A_1 \cup A_2) \leq k_1 + k_2$.

(d) Let $\text{ind}\,(A) = k < \infty$ and consider $\Phi \in C(A, \mathbb{C}^k\backslash\{0\})$ and $n \in \mathbb{N}\backslash\{0\}$ satisfying (1.1). As above, since $0 \notin \Psi(A) = \Phi(A)$ and A is compact by assumption, there exists $\delta > 0$ such that $0 \notin \Psi(A_\delta)$, where $A_\delta := \{u \in X \mid \text{dist}(u, A) \leq \delta\}$. Then, the set $N = A_\delta$ is a closed neighborhood of A, which is invariant, and $\text{ind}\,(N) \leq k$ by construction. And, since the inclusion $A \subset N$ is an equivariant mapping, it follows from property *(b)* that $k = \text{ind}\,(A) \leq \text{ind}\,(N)$. Therefore, $\text{ind}\,(N) = \text{ind}\,(A) = k$. □

Next, let us denote by $\text{Fix}\,(S^1)$ the (closed) invariant subspace of all elements in X which are *fixed* by the representation $\{T(\theta)\}$, that is,

$$\text{Fix}\,(S^1) = \{u \in X \mid T(\theta)u = u \ \ \forall \theta \in S^1\} \ .$$

So, $u \in \text{Fix}\,(S^1)$ if and only if the S^1-orbit $\mathcal{O}(u)$ consists of the singleton $\{u\}$. In this case, we necessarily have that $\text{ind}\,(u) = \infty$ since (1.1) cannot be verified with $\Phi(u) \neq 0$. Indeed, in this case (1.1) becomes

$$\Phi(u) = e^{in\theta}\Phi(u) \quad \forall \theta \in S^1 \ ,$$

hence $\Phi(u) = 0$. On the other hand, one has the following

Proposition 1.2. *([55]) If $u \notin Fix(S^1)$, then* $\text{ind}\,(\mathcal{O}(u)) = 1$.

Proof: Since $u \notin \mathrm{Fix}\,(S^1)$, the continuous mapping $\mathbb{R} \ni \theta \mapsto T(\theta)u \in X$ has minimal period $T > 0$, i.e., $T = \frac{2\pi}{n}$ for some $n \in \mathbb{N}\backslash\{0\}$. Therefore, if we define $\Phi \in C(\mathcal{O}(u), \mathbb{C}\backslash\{0\})$ by the formula

$$\Phi(T(\theta)u) = e^{in\theta} ,$$

we see that $\mathrm{ind}\,(\mathcal{O}(u)) = 1$. □

Remark 1.1. It follows from the above proposition that if $A = \mathcal{O}(u_1) \cup \cdots \cup \mathcal{O}(u_k)$ with $u_j \notin Fix(S^1)$ $\forall j = 1, \ldots k$, then $ind(A) = 1$. Therefore, if $A \in \mathcal{A}$ is such that $A \cap Fix(S^1) = \emptyset$ and $ind(A) \geq 2$, then the set A contains necessarily infinitely many orbits (compare with Proposition 8.2.2 (iii), where $Fix(\mathbb{Z}_2) = \{0\}$).

Remark 1.2. If $Y \subset X$ is a closed, invariant subspace with a topological complement, then Y possesses a topological complement which is also invariant. Indeed, it is enough to take $Z = Q(X)$ where

$$Qu = \frac{1}{2\pi} \int_0^{2\pi} T(-\theta)(I - P)T(\theta)\, u\, d\theta , \quad u \in X ,$$

and $P : X \longrightarrow Y$ is the projection onto Y.

In the next section we will prove a multiplicity result for the case of the above S^1-index. For that, we will need other properties of this S^1-index, whose detailed proofs can be found in [55, 41] and shall be omitted here. Let us then present the necessary properties.

Suppose $Z \subset X$ is a *finite-dimensional, invariant* subspace such that $Z \cap \mathrm{Fix}\,(S^1) = \{0\}$. Then, $\dim Z = 2N$ is even and there exists an isomorphism $J : Z \longrightarrow \mathbb{C}^N$ such that, in Z, $T(\theta)$ is given by the formula

$$T(\theta) = J^{-1} \circ \hat{T}(\theta) \circ J , \tag{1.2}$$

where

$$\hat{T}(\theta)\zeta = (e^{ik_1\theta}\zeta_1, \ldots , e^{ik_N\theta}\zeta_N) \tag{1.3}$$

for all $\theta \in S^1$, $\zeta = (\zeta_1, \ldots , \zeta_N) \in \mathbb{C}^N$ and some $k_j \in \mathbb{N}\backslash\{0\}$, $j = 1, \ldots , N$.

Proposition 1.3. *(cf. [55]) (i) Let* $Y \subset X$ *be a finite-dimensional, invariant subspace and let* $A \subset X$ *be a closed, invariant subspace. If* $Fix(S^1) \subset Y$ *and* $A \cap Y = \emptyset$*, then* codim Y *is even and*

$$\text{ind}\,(A) \le \frac{1}{2}\text{codim } Y \ ;$$

(ii) Let $Z \subset X$ be a finite-dimensional, invariant subspace and let $D \subset Z$ be an open neighborhood of 0 which is invariant. If $Z \cap Fix(S^1) = \{0\}$, then dim Z *is even and*

$$\text{ind}\,(\partial D) = \frac{1}{2}\dim\ Z\ .$$

Proof: *(i)* Let Z be an invariant, topological complement of Y (cf. Remark 1.2). Then dim $Z = 2N$ and $T(\theta)$ is defined on Z by (1.2) and (1.3). Consider the continuous mapping $\Phi = \hat{\Phi} \circ J : Z\backslash\{0\} \longrightarrow \mathbb{C}^N\backslash\{0\}$, where

$$\hat{\Phi}(\zeta) = (\zeta_1^{\frac{n}{k_1}}, \dots, \zeta_N^{\frac{n}{k_N}})\ , \quad \zeta \in \mathbb{C}^N\backslash\{0\}\ ,$$

and n is the minimum common multiple of $k_1, \dots, k_N$ given in (1.3). Then, in view of (1.2) and (1.3), we have that Φ satisfies property (1.1). Therefore,

$$\text{ind}\,(Z\backslash\{0\}) \le \frac{1}{2}\dim\ Z = \frac{1}{2}\text{codim } Y\ . \tag{1.4}$$

On the other hand, since $A \cap Y = \emptyset$, we obtain $P(A) \subset Z\backslash\{0\}$ where $P : X \longrightarrow Z$ is the projection onto Z along Y. Therefore, since P is equivariant (because Y and Z are invariant). Property (b) in the definition of a G-index (Section 8.2) implies that

$$\text{ind}\,(A) \le \text{ind}\,(Z\backslash\{0\})\ . \tag{1.5}$$

Combining (1.4) and (1.5) we obtain $\text{ind}\,(A) \le \frac{1}{2}\text{codim } Y$.

(ii) Let Y be an invariant, topological complement of Z. Then, since $Z \cap \text{Fix}\,(S^1) = \{0\}$, we obtain $\text{Fix}\,(S^1) \subset Y$. Moreover, we have $\partial D \cap Y = \emptyset$. Therefore, part (i) above implies that $\text{ind}\,(\partial D) \le \frac{1}{2}\text{codim } Y = \frac{1}{2}\dim\ Z$. On the other hand, we have $\text{ind}\,(\partial D) \ge \frac{1}{2}\dim\ Z$ (cf. [55]). Thus, $\text{ind}\,(\partial D) = \frac{1}{2}\dim\ Z$. □

2 A Multiplicity Result

We now present a multiplicity result in the case of an S^1-index (cf. [30]).

Theorem 2.1. *Let $\varphi \in C^1(X, \mathbb{R})$ be an invariant functional satisfying the Palais–Smale condition (PS). Suppose $Y, Z \subset X$ are closed, invariant subspaces such that* codim $Y <$ dim Z *and*

(i) $\inf_Y \varphi = a > -\infty$;
(ii) $\sup_{\partial D_r} \varphi = b < +\infty$ *for some* $r > 0$, *where* $\partial D_r = \{u \in Z \mid ||u|| = r\}$;
(iii) $\varphi(u) > b$ *whenever* $\varphi'(u) = 0$, $u \in \mathrm{Fix}\,(S^1)$;
(iv) $\mathrm{Fix}\,(S^1) \subset Y$, $Z \cap \mathrm{Fix}\,(S^1) - \{0\}$

Then, there exist at least $m = \frac{1}{2}(\dim\ Z - \mathrm{codim}\ Y)$ *distinct orbits of critical points of* φ *outside* $\mathrm{Fix}\,(S^1)$ *with critical values in the interval* $[a, b]$.

Proof: Let $p = \frac{1}{2}\mathrm{codim}\ Y$. If $j \geq p + 1$ then (i) implies

$$-\infty < a \leq c_j = \inf_{A \in \mathcal{A}_j} \sup_A \varphi \ , \tag{2.1}$$

since $A \cap Y \neq \emptyset$ for any $A \in \mathcal{A}_j$ (Indeed, if $A \in \mathcal{A}_j$ is such that $A \cap Y = \emptyset$ then (iv) and Proposition 1.3 (i) imply that $\mathrm{ind}\,(A) \leq \frac{1}{2}\mathrm{codim}\ Y = p$, hence $j \leq \mathrm{ind}\,(A) \leq p$).

Let $q := \frac{1}{2}\dim\ Z$ and consider the open ball in Z, $D_r = \{u \in Z \mid ||u|| < r\}$, so that $\mathrm{ind}\,(\partial D_r) = q$, in view of (iv) and Proposition 1.3 (ii). If $j \leq q$, then (ii) implies

$$c_j = \inf_{A \in \mathcal{A}_j} \sup_A \varphi \leq \sup_{\partial D_r} \varphi = b < +\infty \ . \tag{2.2}$$

Therefore, combining (2.1) and (2.2) gives $a \leq c_j \leq b$ for each $j = p + 1, \ldots, q$, that is,

$$-\infty < a \leq c_{p+1} \leq \cdots \leq c_q \leq b < +\infty \ .$$

It follows from Theorem 8.3.5 that each c_j $(j = p+1, \ldots, q)$ is a critical value of φ. If all of these $c_j's$ are distinct we obtain at least $m = q - p$ distinct orbits of critical points of φ, which lie outside $\mathrm{Fix}\,(S^1)$ in view of (iii). On the other hand, if $c_i = c_j$ for some $p + 1 \leq i < j \leq q$, then $\mathrm{ind}\,(K_c) \geq j - i + 1 \geq 2$ and, since $K_c \cap \mathrm{Fix}\,(S^1) = \emptyset$ by (iii), we conclude from Remark 1.1 that K_c has infinitely many orbits in this case. □

Remark 2.1. The first result of the above type is due to Clark [25], where the group $\mathbb{Z}_2$ is considered (cf. Theorem 8.3.6). When $Y = X$ and $\mathrm{Fix}(S^1) = \{0\}$, the above Theorem 2.1 is due to Ekeland and Lasry [38]. Also see [11] for situations in which $\dim\ \mathrm{Fix}\,(S^1)$ is finite.

Remark 2.2. Since the critical values c_j obtained in Theorem 2.1 belong to the interval $[a, b]$, where $a = \inf_Y \varphi$, $b = \max_{\partial D_r} \varphi$, it suffices that a *local* (PS) condition be satisfied.

3 Application to a Class of Problems

In this section we will apply the multiplicity theorem of the previous section to a class of variational problems (cf. Costa–Willem [30]). The abstract context is the following.

Let V be a closed subspace of the Hilbert space $L^2(\Omega;\mathbb{R}^N)$ and consider the equation

$$Lu = \nabla F(u) \tag{$*$}$$

in V, where $F \in C^1(\mathbb{R}^N,\mathbb{R})$ and $L : D(L)\cap V \longrightarrow V$ is an unbounded selfadjoint operator. Assume the following hypotheses:

The spectrum $\sigma(L)$ of L consists of isolated eigenvalues of finite multiplicity,

$$\cdots < \lambda_{-1} < \lambda_0 = 0 < \lambda_1 < \cdots , \tag{h_1}$$

and the corresponding eigenfunctions belong to $L^\infty(\Omega;\mathbb{R}^N)$;

There exist $0 \le \lambda_n < \gamma < \lambda_{n+1}$, $\alpha > 0$ and $0 < \delta < \min\{\gamma-\lambda_n, \lambda_{n+1}-\gamma\}$ such that

$$|\nabla F(u) - \gamma u| \le \delta|u| + \alpha ; \tag{h_2}$$

F is strictly convex, $F(0) = 0$, $\nabla F(0) = 0$ and $\nabla F(u) \in V$ for all $u \in V$; (h_3)

[Note that, from the arguments used in Example C of Chapter 2.1, conditions (h_2) and (h_3) imply that ∇F maps V continuously into V.]

$\nabla F : V \longrightarrow V$ and $L : D(L)\cap V \longrightarrow V$ are equivariant under the isometric representation $\{T(\theta)\}$ of S^1, that is, $\nabla F(T(\theta)u) = T(\theta)\nabla F(u)$,

$$T(\theta)D(L) \subset D(L) \text{ and } LT(\theta)u = T(\theta)Lu. \tag{h_4}$$

Note that (h_4) implies that $\mathrm{Fix}\,(S^1)$ is an invariant subspace under the operator L and the restriction L_0 of L to $\mathrm{Fix}\,(S^1)$ is an equivariant selfadjoint operator with spectrum $\sigma(L_0) \subset \sigma(L)$.

Theorem 3.1. *([30]) Assume the above hypotheses together with*

$\liminf_{|u|\to 0} \frac{2F(u)}{|u|^2} > \lambda_{n+k} \ge \lambda_{n+1}$; (h_5)

$\{\lambda_1, \dots, \lambda_{n+k}\} \cap \sigma(L_0) = \emptyset$; (h_6)

$u = 0$ is the only solution of $(*)$ in $Fix(S^1)$. (h_7)

Then, there exist at least

$$m = \frac{1}{2}\sum_{j=n+1}^{n+k} \dim\ N(L-\lambda_j)$$

distinct orbits of solutions of $(*)$ outside $Fix(S^1)$.

Remark 3.1. If $\sigma(L_0) \subset (-\infty, 0]$, then (h_6) is satisfied in view of (h_1). Moreover, (h_3) implies (h_7) since, in this case, $\nabla F - L_0$ is strictly monotone on $D(L_0)$.

For the proof of Theorem 3.1 we will need some preliminary facts.

Let X be the orthogonal complement of $N(L)$ in V. From (h_1), the operator $-L : D(L) \cap X \longrightarrow X$ possesses a *compact* inverse $K : X \longrightarrow X$.

On the other hand, since $F(0) = 0$, we have $F(u) = \int_0^1 (\nabla F(tu), u)\, dt$, and it follows from (h_2) that, given $\delta < \delta_1 < \min\{\gamma - \lambda_n, \lambda_{n+1} - \gamma\}$, there exists $\alpha_1 > 0$ such that

$$(\gamma - \delta_1)\frac{|u|^2}{2} - \alpha_1 \leq F(u) \leq (\gamma + \delta_1)\frac{|u|^2}{2} + \alpha_1 \quad \forall u \in \mathbb{R}^N.$$

Therefore, in view of (h_3) and Remark 7.1.2, the Legendre–Fenchel transform $G = F^*$ is a strictly convex function of class C^1 satisfying

$$\frac{1}{\gamma + \delta_1}\frac{|v|^2}{2} - \alpha_1 \leq G(v) \leq \frac{1}{\gamma - \delta_1}\frac{|v|^2}{2} + \alpha_1 \quad \forall v \in \mathbb{R}^N. \tag{3.1}$$

And, by Proposition 7.3.1, we can define the dual action $\varphi \in C^1(X, \mathbb{R})$ of Clarke–Ekeland by the formula

$$\varphi(v) = \frac{1}{2}(Kv, v)_V + \int_\Omega G(v)\, dx\ .$$

Lemma 3.2. *(i) If $v \in X$ is a critical point of φ, then $u = \nabla G(v)$ is a solution of $(*)$. Moreover, if $u_1 = \nabla G(v_1)$ and $u_2 = \nabla G(v_2)$ belong to the same orbit, then so do v_1 and v_2;*
(ii) φ satisfies the Palais–Smale condition (PS);
(iii) There exist $\lambda^* > \lambda_{n+k}$ and $\epsilon > 0$ such that

$$G(v) \leq \frac{1}{\lambda^*}\frac{|v|^2}{2}$$

for all $v \in \mathbb{R}^N$ with $|v| \leq \epsilon$.

Proof: *(i)* We have already seen in Section 7.3 that, if $v \in X$ is a critical point of φ, then $u = \nabla G(v) \in D(L)$ is a solution of $(*)$ (cf. Remark 7.3.2). Now, if u_1 and u_2 belong to the same orbit, then $u_2 = T(\theta)u_1$ for some $\theta \in S^1$ and, therefore, in view of (h_4) and the fact that $\nabla G = (\nabla F)^{-1}$, we obtain

$$v_2 = \nabla F(u_2) = \nabla F(T(\theta)u_1) = T(\theta)\nabla F(u_1) = T(\theta)v_1\ ,$$

that is, v_1 and v_2 belong to the same orbit;

(ii) Let $(v_m) \subset X$ be such that $\varphi'(v_m) \to 0$, that is,

$$Kv_m + \nabla G(v_m) - P\nabla G(v_m) = f_m \longrightarrow 0 \text{ in } X,$$

where P is the orthogonal projection onto $N(L)$. Defining $u_m = P\nabla G(v_m) - Kv_m$, we obtain by duality that $v_m = \nabla F(u_m + f_m)$, or $Lu_m = \nabla F(u_m + f_m)$, since $Lu_m = v_m$. Now, assumption (h_2) implies

$$||u_m||_V \leq ||(L-\gamma)^{-1}||\{\delta(||u_m||_V + ||f_m||_V) + \alpha|\Omega| + \gamma||f_m||_V\}$$

where $||(L-\gamma)^{-1}|| = \max\{(\gamma - \lambda_n)^{-1}, (\lambda_{n+1} - \gamma)^{-1}\} < \delta^{-1}$. Therefore, since $||f_m||_V \to 0$, we obtain that $||u_m||_V$ is bounded. In view of the linear growth of ∇F and ∇G (cf. (3.3) in the proof of Proposition 7.3.1), we conclude that $v_m = \nabla F(u_m + f_m)$ and $\nabla G(v_m)$ are also bounded in V. Therefore, passing to a subsequence if necessary, we may assume that $v_m \rightharpoonup v$ weakly in X and $P\nabla G(v_m) \to w \in N(L)$, since $N(L)$ is finite dimensional by (h_1). Since K is a compact operator, we obtain that $Kv_m \to Kv$ in X, so that

$$v_m = \nabla F(P\nabla G(v_m) - Kv_m + f_m) \longrightarrow \nabla F(w - Kv) \quad \text{in } X.$$

Therefore, $v_m \to v$ in X.

(iii) The proof follows from assumption (h_5) and was already done during the proof of Lemma 7.4.3. Indeed, (h_5) implies that there exist $\delta > 0$ and $\lambda^* > \lambda_{n+k}$ such that $|F(u)| \geq \frac{1}{2}\lambda^*|u|^2$ for $|u| \leq \delta$. On the other hand, there exists $\epsilon > 0$ such that $|\nabla G(v)| \leq \delta$ for $|v| \leq \epsilon$. Therefore, since $G(v) = (u, v) - F(u)$ with $u = \nabla G(v)$, we obtain for $|v| \leq \epsilon$ that

$$\begin{aligned} G(v) &\leq \max_{|u|\leq\delta}\{(u,v) - \tfrac{1}{2}\lambda^*|u|^2\} \\ &\leq \max_u\{(u,v) - \tfrac{1}{2}\lambda^*|u|^2\} = \tfrac{1}{\lambda^*}\tfrac{|v|^2}{2}\,. \end{aligned}$$

□

Proof of Theorem 3.1: We will apply Theorem 2.1 to the dual action $\varphi \in C^1(X, \mathbb{R})$, which is invariant by assumption (h_4) and satisfies (PS) by Lemma 3.2(ii) above. Let us define the following subspaces of X:

$$Y = [N(L - \lambda_0) \oplus \cdots \oplus N(L - \lambda_n)]^\perp\,,$$

$$Z = N(L - \lambda_1) \oplus \cdots \oplus N(L - \lambda_{n+k})\,.$$

Then, we have $\mathrm{Fix}\,(S^1) \cap X \subset Y$ and $Z \cap \mathrm{Fix}\,(S^1) = \{0\}$ in view of assumption (h_6), so that condition (iv) of Theorem 2.1 is satisfied.

Now, through eigenfunction expansion of the elements of Y, we obtain the estimate

$$(Kv, v)_V \geq -\frac{1}{\lambda_{n+1}}||v||_V^2 \quad \forall v \in Y \ ,$$

which, combined with (3.1), gives

$$\varphi(v) \geq \frac{1}{2}\left(-\frac{1}{\lambda_{n+1}} + \frac{1}{\gamma + \delta_1}\right)||v||_V^2 - \alpha_1|\Omega| \quad \forall v \subset Y.$$

Therefore, since $\gamma + \delta_1 < \lambda_{n+1}$, it follows that φ is bounded from below on Y and condition (i) of Theorem 2.1 is satisfied.

In order to verify condition (ii), we again use eigenfunction expansion to obtain the following estimate in Z,

$$(Kv, v)_V \leq -\frac{1}{\lambda_{n+k}}||v||_V^2 \quad \forall v \in Z \ ,$$

so that Lemma 3.2 (iii) above gives

$$\varphi(v) \leq \frac{1}{2}\left(-\frac{1}{\lambda_{n+k}} + \frac{1}{\lambda^*}\right)||v||_V^2 < 0$$

for all $v \in Z$ with $0 < ||v||_{L^\infty} \leq \epsilon$. Since Z is finite dimensional, there exists $r > 0$ such that $\sup_{\partial D_r} \varphi = b < 0$, where $\partial D_r = \{v \in Z \mid ||v|| = r\}$. Therefore, condition (ii) of Theorem 2.1 is also satisfied.

Finally, if $v \in Fix(S^1)$ is a critical point of φ, then $u = \nabla G(v) \in D(L)$ is a solution of $(*)$ (cf. Lemma 3.2) and, by the equivariance of $\nabla G = (\nabla F)^{-1}$ (assumption (h_4)), it follows that also $u \in \mathrm{Fix}\,(S^1)$. Therefore, (h_7) implies $u = 0$, so that $v = \nabla F(u) = 0$ and $\varphi(v) = 0$. This shows that condition (iii) of Theorem 2.1 is satisfied. Therefore, that theorem gives the existence of at least

$$\frac{1}{2}\,(\dim\ Z - \mathrm{codim}\ Y) = \frac{1}{2}\sum_{j=n+1}^{j=n+k} \dim\ N(L - \lambda_j)$$

distinct orbits of critical points of φ outside $\mathrm{Fix}\,(S^1)$. And these distinct orbits give rise to distinct solutions of $(*)$ outside $\mathrm{Fix}\,(S^1)$ in view of Lemma 3.2 (i). The proof is complete. □

4 A Dirichlet Problem on a Plane Disk

Next we illustrate Theorem 3.1 by considering the Dirichlet problem

$$\begin{cases} -\Delta u = g(u) & \text{in } \Omega \\ \quad\ u = 0 & \text{on } \partial\Omega\ . \end{cases} \tag{4.1}$$

where Ω is the unit disk in $\mathbb{R}^2$. Other applications can be found in [30].

Let A be the operator $-\Delta$ on the space $V = L^2(\Omega, \mathbb{R})$ with Dirichlet boundary condition, that is, $D(A) = H^2(\Omega) \cap H_0^1(\Omega)$ and $Au = -\Delta u \ \forall u \in D(A)$. The eigenvalues of A are of the form $\mu = \nu^2$, where ν is a positive zero of some Bessel function of the first kind, J_n, $n = 0, 1, 2 \dots$ (see Courant–Hilbert [34], Vol. II). If $n \geq 1$, the corresponding eigenfunctions are given in polar coordinates (r, θ) by

$$J_n(\nu r) \cos\, n\theta \ , \quad J_n(\nu r) \sin\, n\theta \ .$$

If ν is a zero of J_0, then $J_0(\nu r)$ is a (radially symmetric) eigenfunction corresponding to the eigenvalue $\mu = \nu^2$. Therefore, if we denote the spectrum of A by $\sigma(A) = \{\mu_1, \mu_2, \dots\}$, with $0 < \mu_1 < \mu_2 < \cdots$, then each eigenvalue is either simple or of even multiplicity: in fact, the nonsimple eigenvalues are double, since a result of C. Siegel (cf. [73], pg. 485) implies that the positive zeros of J_{n_1} and J_{n_2} are distinct if $n_1 \neq n_2$.

Theorem 4.1. *([30]) Let $g : \mathbb{R} \longrightarrow \mathbb{R}$ be continuous with $g(0) = 0$ and assume that*

$$\mu_p < \frac{g(u) - g(v)}{u - v} \quad \forall u \neq v \quad \textit{and} \quad \frac{g(u)}{u} < \mu_q \quad \forall u \neq 0 \ , \tag{4.2}$$

$$\mu_n < \liminf_{|u| \to \infty} \frac{g(u)}{u} \leq \limsup_{|u| \to \infty} \frac{g(u)}{u} < \mu_{n+1} \leq \mu_{n+k} < \liminf_{|u| \to 0} \frac{g(u)}{u} \ , \tag{4.3}$$

where $\mu_p \leq \mu_n, \mu_{n+k} < \mu_q$ and $\{\mu_{p+1}, \dots, \mu_{q-1}\} \cap \{\mu > 0 \mid J_0(\sqrt{\mu}) = 0\} = \emptyset$.[1] *Then, Problem* (4.1) *possesses at least k weak solutions which are non-radially symmetric and geometrically distinct (two solutions are geometric distinct if one is not a rotation of the other).*

Proof: Denote by L the operator $A - \mu_p$ on the space $V = L^2(\Omega, \mathbb{R})$ with Dirichlet boundary condition, and define $F(u) \in C^1(\mathbb{R}, \mathbb{R})$ by $F(u) = \int_0^u g(s)\, ds - \frac{1}{2}\mu_p u^2$ and $f = F'$. Then, assumption (4.2) and the fact that $g(0) = 0$ imply (h_3) and, since $\sigma(L) = \{\mu_j - \mu_p \mid j = 1, 2, \dots\}$, we also obtain (h_2) and (h_5) from (4.3).

On the other hand, consider the isometric representation of S^1 on V defined by the rotations in Ω, that is,

$$(T(\theta)u)(x) = u(R(\theta)x) \ ,$$

[1] It follows that the eigenvalues crossed by $\frac{g(u)}{u}$ are all double.

where $R(\theta)x = R(\theta)(x_1, x_2) = (x_1 \cos\theta - x_2 \sin\theta, x_1 \sin\theta + x_2 \cos\theta)$. Then, it is clear that L and F' are equivariant, so that (h_4) is also satisfied. And, since $\mathrm{Fix}\,(S^1)$ is the space of radially symmetric function, condition (h_6) is satisfied because each $\lambda_j = \mu_{j+p} - \mu_p$, $j = 1, \ldots, q-1-p$, is a double eigenvalue of L.

Finally, in order to verify (h_7), we observe that (4.2) implies that $\frac{f(u)}{u} < \mu_q - \mu_p$ for all $u \neq 0$, that is,

$$f(u)^2 < (\mu_q - \mu_p) f(u) u \;, \quad \forall u \neq 0 \;.$$

Therefore, if $u \neq 0$ is a radially symmetric solution of (4.1), we obtain

$$\begin{aligned} \|f(u)\|_{L^2}^2 &< (\mu_q - \mu_p)(f(u), u)_{L^2} = (\mu_q - \mu_p)(Lu, u)_{L^2} \\ &< \|Lu\|_{L^2}^2 = \|f(u)\|_{L^2}^2 \;. \end{aligned}$$

This absurdity implies that (h_7) is also satisfied.

Therefore, Theorem 3.1 guarantees the existence of at least k nonradially symmetric and geometrically distinct weak solutions of Problem (4.1). □

10

Problems with Lack of Compactness

1 Introduction

In this chapter and the next we will present two examples of situations in which the variational problem under consideration lacks some desirable compactness properties. Typically, lack of compactness is due to the action of a group under which the pertinent functional is invariant. For example, an autonomous semilinear elliptic equation in the whole space $\mathbb{R}^N$,

$$-\Delta u + u = h(u) \ , \tag{1.1}$$

is invariant under the group of translations $u(\cdot) \mapsto u(\cdot + z)$, $z \in \mathbb{R}^N$. We will be considering one such situation in this chapter.

Now, in case Problem (1.1) is posed in a bounded domain $\Omega \subset \mathbb{R}^N$, $N \geq 3$, a lack of compactness can happen when, for example, the nonlinearity $h(u)$ has a *critical growth* at the limiting exponent $q = 2^* - 1 := (N+2)/(N-2)$ for the Sobolev embedding $H^1(\mathbb{R}^N) \hookrightarrow L^{2^*}(\mathbb{R}^N)$.

Both situations and generalizations have been extensively considered in the literature through the use of the *concentration-compactness method* of P.-L. Lions [51, 52]. The reader is invited to learn how this method is used in the above original papers. In this chapter, when considering our example in the whole space $\mathbb{R}^N$, we shall neither use the concentration-compactness method per se, nor shall we be concerned with making the best possible assumptions on h. Instead, we will impose additional restrictions on the nonlinearity h which will allow us to consider the associated *Nehari set*

$$M = \{u \in H^1(\mathbb{R}^N) \mid \int_{\mathbb{R}^N} (|\nabla u|^2 + u^2)\, dx = \int_{\mathbb{R}^N} h(u) u \, dx \ \}$$

as a C^1-submanifold of $H^1(\mathbb{R}^N)$ (of codimension 1), which is a natural constraint for our functional in the sense that the critical points of the restriction

$\varphi|M$ are precisely the critical points of φ. In particular, we can find a solution of Problem (1.1) via *global minimization* on M (as in Chapter 6). Such a solution is called a *ground-state solution* since it has the least possible nonzero energy $\varphi(u)$.

We point out that, in fact, all that is needed in order to obtain a solution of Problem (1.1) through *constrained minimization* is to have h subcritical satisfying $H(\zeta) > \frac{1}{2}\zeta^2$ for some $\zeta > 0$ (cf. the original paper [14, Part I] of Berestycki and Lions). Under these minimal assumptions, a solution of (1.1) is found in [14] through the *constrained minimization* problem of finding the *global minimum* of

$$\int_{\mathbb{R}^N} |\nabla u|^2 \, dx$$

restricted to

$$\{u \in H^1(\mathbb{R}^N) \mid \int_{\mathbb{R}^N} (H(u) - \frac{1}{2}u^2) \, dx = 1 \}.$$

We start with preliminary results involving two beautiful lemmas due to Brézis and Lieb [19] and to Lions [51], respectively, and which deal with the question of when a weakly convergent L^p-sequence is strongly convergent. As usual, in what follows we will be denoting by 2^* the limiting exponent in the Sobolev embedding $H^1(\mathbb{R}^N) \hookrightarrow L^p(\mathbb{R}^N)$, $2 \leq p \leq 2^*$, namely, $2^* = 2N/(N-2)$ if $N \geq 3$ and $2^* = \infty$ if $N = 1, 2$.

2 Two Beautiful Lemmas

As is well known, given a domain $\Omega \subset \mathbb{R}^N$ and a bounded sequence (u_n) of functions in $L^p(\Omega)$ $(1 < p < \infty)$ one can extract a subsequence (still denoted by u_n, for simplicity) which converges weakly and pointwisely to some $u \in L^p(\Omega)$. In addition the Fatou lemma gives

$$\int_\Omega |u|^p \leq \liminf_{n\to\infty} \int_\Omega |u_n|^p \ ,$$

with equality occurring if and only if u_n converges strongly to u in $L^p(\Omega)$. As it turns out, Brézis and Lieb obtained a refinement of the Fatou lemma in [19] by showing that the difference $||u_n||^p_{L^p} - ||u||^p_{L^p}$ behaves exactly like $||u_n - u||^p_{L^p}$ as $n \to \infty$. In other words, they found the *missing term* in the inequality of the Fatou lemma by proving that

$$||u||^p_{L^p} = \lim_{n\to\infty} (||u_n||^p_{L^p} - ||u_n - u||^p_{L^p}) \ .$$

Their proof essentially uses the fact that the continuous function $H(s) := |s|^p$ satisfies the following property:

$$\forall \epsilon > 0 \; \exists C_\epsilon > 0 \text{ with } |H(s+t) - H(s)| \leq \epsilon H(s) + C_\epsilon H(t) \;\; \forall s, t \in \mathbb{R} \; . \tag{2.1}$$

Indeed, as mentioned in [19], the result holds for more general functions and we shall state it here in a more general form.

Brézis–Lieb Lemma. *Let $H : \mathbb{R} \longrightarrow [0, \infty)$ be continuous and satisfy (2.1). Suppose a sequence (u_n) of measurable functions in Ω satisfies $u_n \to u$ a.e., $\sup_n \int_\Omega H(u_n) < \infty$ and $\int_\Omega H(u) < \infty$. Then $\sup_n \int_\Omega H(u_n - u) < \infty$ and*

$$\int_\Omega |H(u_n) - H(u) - H(u_n - u)| \longrightarrow 0 \quad \text{as } n \to \infty.$$

Proof: We first note that when (u_n) is a bounded sequence in $L^p(\Omega)$ the above conditions are automatically satisfied with $H(s) := |s|^p$ (and for a suitable subsequence of (u_n)).

In view of condition (2.1) one has the estimate

$$|H(u_n) - H(u_n - u)| \leq \epsilon H(u_n - u) + C_\epsilon H(u) \; ,$$

so that, with $\epsilon = 1/2$,

$$H(u_n - u) \leq 2 \left[H(u_n) + C_{1/2} H(u) \right]$$

which shows that $K := \sup_n \int_\Omega H(u_n - u) < \infty$ in view of the assumptions. Moreover, one has the estimate

$$h_{n,\epsilon} := \left(|H(u_n) - H(u) - H(u_n - u)| - \epsilon H(u_n - u) \right)^+ \leq (1 + C_\epsilon) \, H(u),$$

so that $\int_\Omega h_{n,\epsilon} \to 0$ as $n \to \infty$ by the Lebesgue theorem. Therefore, since

$$|H(u_n) - H(u) - H(u_n - u)| \leq h_{n,\epsilon} + \epsilon H(u_n - u) \; ,$$

we obtain

$$\limsup_{n \to \infty} \int_\Omega |H(u_n) - H(u) - H(u_n - u)| \leq K\epsilon \; .$$

The result follows since $\epsilon > 0$ was arbitrary. □

Remark 2.1. As pointed out in [19], condition (2.1) is satisfied by general *convex* functions. On the other hand, let $H : \mathbb{R} \longrightarrow \mathbb{R}$ be continuous and satisfy $H(s) > 0$ for $s \neq 0$. And suppose that $H(s)$ behaves asymptotically as pure powers as $|s| \to 0$ and as $|s| \to \infty$ in the sense that, for some $0 < p, q < \infty$, one has

$$\lim_{|s|\to 0} \frac{H(s)}{|s|^p} = l_0 > 0 \quad \text{and} \quad \lim_{|s|\to\infty} \frac{H(s)}{|s|^q} = l_\infty > 0 \; .$$

Then it can be shown (cf. [32]) that $H(s)$ also satisfies condition (2.1).

Next, we consider a lemma by Lions [51] which gives conditions guaranteeing strong L^p-convergence of a weakly convergent L^p-sequence arising from a bounded H^1-sequence of functions.

Lions Lemma. *Suppose $u_n \in H^1(\mathbb{R}^N)$ is a bounded sequence satisfying*

$$\lim_{n\to\infty} \left(\sup_{y\in\mathbb{R}^N} \int_{B_r(y)} |u_n|^p \, dx \right) = 0$$

for some $p \in [2, 2^*)$ and $r > 0$, where $B_r(y)$ denotes the open ball of radius r centered at $y \in \mathbb{R}^N$. Then, $u_n \longrightarrow 0$ strongly in $L^q(\mathbb{R}^N)$ for all $2 < q < 2^*$.

Proof: We only consider the case $N \geq 3$ and reproduce the proof as given in [75]. Let $p < s < 2^*$ be given and consider $\theta \in (0,1)$ defined by

$$\frac{1}{s} = \frac{1-\theta}{p} + \frac{\theta}{2^*} \; .$$

Then, writing $\hat{B} = B_r(y)$ for short and using Hölder's inequality together with the Sobolev embedding, we have

$$\begin{aligned} ||u||_{L^s(\hat{B})} &\leq ||u||^{1-\theta}_{L^p(\hat{B})} ||u||^{\theta}_{L^{2^*}(\hat{B})} \\ &\leq C||u||^{1-\theta}_{L^p(\hat{B})} \left(\textstyle\int_{\hat{B}} (|u|^2 + |\nabla u|^2) \, dx \right)^{\frac{\theta}{2}} \; , \end{aligned}$$

hence

$$\int_{\hat{B}} |u|^s \, dx \leq C^s ||u||^{s(1-\theta)}_{L^p(\hat{B})} \int_{\hat{B}} (|u|^2 + |\nabla u|^2) \, dx$$

by picking $\theta = \frac{2}{s}$. Next, if we consider a covering of $\mathbb{R}^N$ by balls $\hat{B} = B_r(y)$ of radius r as above, so that each point in $\mathbb{R}^N$ is contained in $(N+1)$ such balls (at most), we obtain the following estimate

$$\int_{\mathbb{R}^N} |u|^s \, dx \leq (N+1) C^s \sup_{y\in\mathbb{R}^N} \left(\int_{B_r(y)} |u|^p \, dx \right)^{\frac{s(1-\theta)}{p}} \int_{\mathbb{R}^N} (|u|^2 + |\nabla u|^2) \, dx \; ,$$

which shows by the given assumption that $u_m \longrightarrow 0$ in $L^s(\mathbb{R}^N)$. Therefore, since $2 < s < 2^*$, we conclude (again by the Hölder and Sobolev inequalities) that $u_m \longrightarrow 0$ in $L^q(\mathbb{R}^N)$ for $2 < q < 2^*$. □

3 A Problem in $\mathbb{R}^N$

As a first application of a problem lacking compactness, we consider a subcritical elliptic problem in all of $\mathbb{R}^N$:

$$\Delta u + u = h(u) \quad , \quad u \in H^1(\mathbb{R}^N) \, . \tag{3.1}$$

In the literature, such an equation is called a *field equation* in the *positive mass case* (because of the term u in the left-hand side of the equation), an important example being that when $h(u)$ is a pure power. Below we list some of the assumptions on the nonlinearity $h(s)$ that will be made in the course of this section, where we will always be assuming that $N \geq 3$. As usual, we denote $H(s) = \int_0^s h(t)\, dt$ and, for simplicity of notation, all integrals are understood as taken over all of $\mathbb{R}^N$, unless otherwise stated:

$h \in C^1(\mathbb{R})$ satisfies $h(0) = 0$, $\lim_{|s|\to 0} \frac{|h'(s)|}{|s|^\delta} < +\infty$ for some $\delta > 0$, and

$$\limsup_{|s|\to\infty} \frac{|h'(s)|}{|s|^{p-2}} < +\infty \text{ for some } 2 < p < 2^* \, ; \tag{h_1}$$

$H(s)$ and $[sh(s) - 2H(s)]$ are *strictly convex* functions with

$$\lim_{|s|\to\infty} \frac{H(s)}{s^2} = +\infty \, . \tag{h_2}$$

Remark 3.1. In view of (h_1) we note that the functional

$$\varphi(u) = \frac{1}{2}\int(|\nabla u|^2 + u^2)\, dx - \int H(u)\, dx = \frac{1}{2}\|u\|^2 - \int H(u)\, dx$$

is well defined and of class C^2 on $H^1(\mathbb{R}^N)$, with its critical points being precisely the solutions of (3.1). Moreover, Problem (3.1) has the trivial solution $u = 0$ and lacks compactness in the sense that, for any nontrivial solution v, the sequence $v_n(x) := v(x + z_n)$ with $z_n \in \mathbb{R}^N$ satisfying $|z_n| \to \infty$ (so that $\|v_n\| = \|v\|$ and $v_n \rightharpoonup 0$ weakly in $H^1(\mathbb{R}^N)$) has the property that

$$\varphi(v_n) = \varphi(v) \quad , \quad \varphi'(v_n) = 0 \, ,$$

but no subsequence of v_n converges strongly in $H^1(\mathbb{R}^N)$. In particular, φ does not satisfy $(PS)_c$ for $c = \varphi(v)$.

Remark 3.2. Since we are assuming $h(0) = 0$ it is not hard to see that (h_2) implies (in fact, is equivalent to):

$(\hat{h}_2)$ $h(s)$ is *strictly increasing* with $\lim_{|s|\to\infty} H(s)/s^2 = +\infty$ and

$$h'(s) > \frac{h(s)}{s} > \frac{2H(s)}{s^2} > 0 \quad \forall s \neq 0 \ .$$

In particular, any such h is *superlinear* in the sense that $\lim_{|s|\to\infty} h(s)/s = +\infty$ (in fact, $\lim_{|s|\to\infty} h'(s) = +\infty$), and we also have that

$$h'(s)s^2 - h(s)s > 0 \quad \text{if } s \neq 0.$$

Remark 3.3. A typical example of a function satisfying (h_1) and (h_2) is the pure power $s \mapsto a|s|^{p-2}s$, with $2 < p < 2^*$ and some $a > 0$ or, more generally, the function

$$h_{p,q}(s) := \begin{cases} a|s|^{p-2}s \ if \ s \geq 0 \\ b|s|^{q-2}s \ if \ s < 0 \end{cases}$$

with $2 < p, q < 2^*$ and $a, b > 0$. Still another example of a function (with a much slower growth) but satisfying the same conditions above is given by

$$h(s) = H'(s) \ , \quad H(s) = s^2 ln \ (1 + s^2) \ .$$

On the other hand, the first two examples satisfy the stronger superlinear Ambrosetti–Rabinowitz (global) condition

$$sh(s) \geq \theta H(s) > 0 \ \text{ for } s \neq 0 \text{ (and some } \theta > 2)$$

whereas the last one does not.

Under some suitably mild condition on h (say h continuous with $|h(s)| \leq A|s| + B|s|^{p-1}$ for some $A, B > 0$ and $2 < p \leq 2N/(N-2)$ if $N \geq 3$), we can define the so-called *Nehari set*

$$M := \{ \ u \in H^1(\mathbb{R}^N)\backslash\{0\} \mid \psi(u) := \|u\|^2 - \int h(u)u \ dx = 0 \ \} \ .$$

Of course, we may have $M = \emptyset$. However, if $u \in H^1(\mathbb{R}^N)$ is a nonzero weak solution of (3.1) then $u \in M$.

Lemma 3.1. *Assume* (h_1). *If* $\lim_{|s|\to\infty} h(s)/s = +\infty$ *(h is superlinear) and* $\gamma(s) := sh(s) - 2H(s)$ *is a strictly convex function, then*

(i) $M \neq \emptyset$ *is a* C^1*-submanifold of* $H^1(\mathbb{R}^N)$ *of codimension* 1 *and* $0 \notin \overline{M}$*;*
(ii) M *is a natural constraint for* φ*, that is,* $0 \neq u \in H^1(\mathbb{R}^N)$ *is a critical point of* φ *if and only if* $u \in M$ *and* u *is a critical point of* $\varphi|M$*.*

Proof: (*i*) In view of (h_1), the functional ψ is clearly of class C^2 on $H^1(\mathbb{R}^N)$. Let $0 \neq v \in H^1(\mathbb{R}^N)$ and consider the function $0 < t \mapsto \psi(tv)$. Again, in view of (h_1) and Sobolev embedding we claim that $\psi(tv) > 0$ for $t > 0$ small. Indeed, we have

$$\begin{aligned} |\int h(tv)tv\, dx| &\leq \epsilon t^2 \int |v|^2\, dx + \frac{1}{p-1} C_\epsilon |t|^p \int |v|^p\, dx \\ &\leq \epsilon t^2 \|v\|^2 + D_\epsilon |t|^p \|v\|^p \end{aligned}$$

so that, by picking $0 < \epsilon < 1$, we obtain

$$t^2\|v\|^2 - \int h(tv)tv\, dx \geq t^2(1-\epsilon)\|v\|^2 - |t|^p D_\epsilon \|v\|^p > 0$$

for all $t > 0$ sufficiently small.

Similarly, it is not hard to see that the superlinearity of h together with the fact that $h(s)s \geq 0$ imply that $\lim_{t\to\infty} \psi(tv) = -\infty$. Therefore, there exists $\bar{t}$ such that $\bar{t}v \in M$. In particular, $M \neq \emptyset$.

Next, from the strict convexity of $\gamma(s) = sh(s) - 2H(s)$ we have

$$\gamma'(s) = sh'(s) - h(s) < 0 \quad \text{if } s < 0\ ,$$

$$\gamma'(s) = sh'(s) - h(s) > 0 \quad \text{if } s > 0\ ,$$

hence

$$h'(s) > \frac{h(s)}{s} \quad \forall s \neq 0\ . \tag{3.2}$$

We claim that $\psi'(u) \neq 0$ for all $u \in M$. Indeed, if $u \in M$ is such that

$$\psi'(u) \cdot w = 2\langle u, w\rangle - \int h(u)w\, dx - \int h'(u)uw\, dx = 0$$

for all $w \in H^1(\mathbb{R}^N)$, then, by picking $w = u$, it follows that

$$2\|u\|^2 - \int h(u)u\, dx - \int h'(u)u^2\, dx = 0\ ,$$

so that

$$2\|u\|^2 > 2\int h(u)u\, dx\ ,$$

in view of (3.2). This is in contradiction with the fact that $u \in M$. Therefore, M is a C^1-submanifold of $H^1(\mathbb{R}^N)$ of codimension 1.

We now show that $0 \notin \overline{M}$. As before, (h_1) and Sobolev embedding imply, for $u \in M$ and $\epsilon > 0$,

$$\begin{aligned}\|u\|^2 &= \int h(u)u \, dx \leq \epsilon \int |u|^2 \, dx + \frac{1}{p-1} C_\epsilon \int |u|^p \, dx \\ &\leq \epsilon \|u\|^2 + D_\epsilon \|u\|^p\end{aligned}$$

where $D_\epsilon > 0$. By taking (say) $\epsilon = 1/2$ and recalling that $u \neq 0$ by definition of M, we obtain

$$\|u\|^{p-2} \geq \frac{1}{2D_{1/2}} > 0 \quad \forall u \in M \ .$$

Therefore, $0 \notin \overline{M}$.

(*ii*) Let $0 \neq u \in H^1(\mathbb{R}^N)$ be a critical point of φ. Then,

$$\varphi'(u) \cdot w = \langle u, w \rangle - \int h(u) w \, dx = 0 \quad \forall w \in H^1(\mathbb{R}^N) \ ,$$

and choosing $w = u$ yields

$$\|u\|^2 - \int h(u) u \, dx = 0 \ ,$$

that is, $u \in M$. Conversely, if $u \in M$ is a critical point of $\varphi|M$, then

$$\varphi'(u) = \lambda \psi'(u)$$

for some Lagrange multiplier $\lambda \in \mathbb{R}$, that is,

$$\langle u, w \rangle - \int h(u) w \, dx = \lambda \left[2\langle u, w \rangle - \int h(u) w \, dx - \int h'(u) u w \, dx \right]$$

for all $w \in H^1(\mathbb{R}^N)$. Once again, choosing $w = u$ and recalling that $u \in M$ we obtain

$$0 = 2\lambda \|u\|^2 - \lambda \left[\int h(u) u \, dx + \int h'(u) u^2 \, dx \right] \ ,$$

or,

$$2\lambda \|u\|^2 = \lambda \left[\int h(u) u \, dx + \int h'(u) u^2 \, dx \right] \ .$$

Since the above expression in square brackets is greater than $2 \int h(u) u \, dx$ by (3.2), we necessarily must have $\lambda = 0$ in order not to contradict the fact that $u \in M$. Thus u is a critical point of φ. □

Lemma 3.2. *Under the assumptions in Lemma 3.1, define*

$$c_* := \inf_{v \in M} \varphi(v) \ .$$

Then, $c_* > 0$ *and* c_* *is the ground-state level for Problem* (3.1), *that is,*

$$c_* = \inf\{ \ \varphi(u) \ | \ \ u \neq 0 \textit{ is a solution of Problem } (3.1) \ \} \ .$$

Proof: We start by observing that the functional φ has mountain-pass geometry, that is,

(i) There exists $a, \delta > 0$ such that $\varphi(u) \geq a$ if $\|u\| = \delta$;
(ii) There exists $v_0 \in H^1(\mathbb{R}^N)$ such that $\|v_0\| > \delta$ and $\varphi(v_0) \leq 0$.

Indeed, as in Lemma 3.1, using (h_1) and Sobolev inequality yields (with $0 < \epsilon < 1$)

$$\varphi(u) = \frac{1}{2}\|u\|^2 - \int H(u)\, dx \geq \frac{1}{2}(1-\epsilon)\|u\|^2 - D_\epsilon \|u\|^p \,.$$

Therefore (i) holds true with $\|u\| = \delta > 0$ sufficiently small so that the right-hand side $a = (1-\epsilon)\delta^2/2 - D_\epsilon \delta^p$ of the above inequality is positive.

Similarly, as in Lemma 3.1, we obtain (ii) from the fact that, for each $0 \neq v \in H^1(\mathbb{R}^N)$, we have

$$\lim_{t\to\infty} \varphi(tv) = -\infty \,.$$

Moreover, as also observed before, the strict convexity of $[h(s)s - 2H(s)]$ gives

$$h'(s) > \frac{h(s)}{s} \quad \forall s \neq 0 \,.$$

This in turn means that the function $s \mapsto h(s)/s$ is strictly decreasing on $(-\infty, 0]$ and strictly increasing on $[0, \infty)$. Therefore, for each fixed $\|w\| = 1$, the real-valued function

$$0 < t \mapsto \varphi(tw) = \frac{1}{2}t^2 - \int H(tw)\, dx$$

has a unique critical point $\hat{t} = \hat{t}(w) > 0$ (its global maximum). And since $\hat{t} = \int h(\hat{t}w)w\, dx$, we can characterize the Nehari manifold as follows:

$$M = \{\ \hat{t}(w)w \mid \|w\| = 1\ \} \,.$$

Next, letting $\Gamma := \{\ \gamma \in C([0,1], H^1(\mathbb{R}^N)) \mid \gamma(0) = 0, \quad \varphi(\gamma(1)) < 0\ \}$, it follows from $(i), (ii)$ that the *mountain-pass level* of φ

$$c_{MP} := \inf_{\gamma\in\Gamma} \sup_{0\leq t\leq 1} \varphi(\gamma(t))$$

is *positive*. In particular,

$$0 < c_{MP} \leq c_* \,.$$

In fact, in our present case, we have the equality

$$c_{MP} = c_*$$

since, by arguments similar to those we have been using so far, it is not hard to show that for any admissible path $\gamma \in \Gamma$ one has $\gamma(\bar{t}) \in M$ for some $0 < \bar{t} < 1$. □

Lemma 3.3. *Assume* (h_1). *Then any minimizing sequence* (u_n) *for* c_* *is bounded.*

Proof: Let $u_n \in M$ be a minimizing sequence for c_*:

$$\|u_n\|^2 - \int h(u_n)u_n \, dx = 0 \, , \tag{3.3}$$

$$\frac{1}{2}\|u_n\|^2 - \int H(u_n) \, dx \longrightarrow c_* \, . \tag{3.4}$$

Claim: There exists $C > 0$ such that $\|u_n\| \le C \; \forall n \in \mathbb{N}$. Suppose by contradiction that (for some subsequence, still labeled u_n for simplicity) we have $t_n := \|u_n\| \to \infty$. Picking $\delta > 2$, and letting $v_n := \sqrt{\delta c_*} u_n / t_n$, it follows that $\|v_n\|^2 = \delta c_*$ and there exists $v \in H^1(\mathbb{R}^N)$ such that

$$\begin{cases} v_n \rightharpoonup v \text{ weakly in } H^1(\mathbb{R}^N) \\ v_n \to v \text{ strongly in } L^q_{loc}(\mathbb{R}^N) \text{ (for any } 2 \le q < 2^*) \\ v_n \to v \text{ a.e. in } \mathbb{R}^N. \end{cases}$$

Sub-Claim: $v \neq 0$.

Indeed, since (v_n) is bounded in $H^1(\mathbb{R}^N)$, we consider the quantity

$$\liminf_{n \to \infty} \left(\sup_{y \in \mathbb{R}^N} \int_{B_1(y)} |v_n|^2 \, dx \right) := \alpha \, . \tag{3.5}$$

If $\alpha = 0$, then Lions lemma says (passing to a subsequence, if necessary) that $v_n \longrightarrow 0$ strongly in $L^q(\mathbb{R}^N)$ for all $2 < q < 2^*$ and, since (h_1) implies

$$\int H(v_n) \, dx \le \frac{1}{2}\epsilon \|v_n\|^2_{L^2} + \frac{1}{p(p-1)} C_\epsilon \|v_n\|^p_{L^p} \, ,$$

we obtain $\int H(v_n) \, dx \le C\epsilon + o(1)$, where $C = \delta c_*/2$. Since $\epsilon > 0$ was arbitrary we conclude that

$$\varphi(v_n) = \frac{1}{2}\|v_n\|^2 - \int H(v_n) \, dx = \frac{\delta}{2} c_* + o(1) \, . \tag{3.6}$$

On the other hand, since u_n belongs to M and $v_n = \tau_n u_n$, we have that

$$\varphi(v_n) \le \varphi(u_n) = c_* + o(1) \, ,$$

which is in contradiction with (3.6) since $\delta > 2$ by our choice. Therefore, we must have $\alpha > 0$ in (3.5), so that

$$\int_{B_1(0)} |\widetilde{v}_n|^2 \, dx \geq \frac{\alpha}{2} > 0 \tag{3.7}$$

for n large, where $\widetilde{v}_n(x) := v_n(x + z_n)$ for some $z_n \in \mathbb{R}^N$, that is,

$$\widetilde{v}_n(x) := v_n(x + z_n) = \frac{\sqrt{\delta c_*}}{t_n} \widetilde{u}_n(x) \quad , \quad \widetilde{u}_n(x) := u_n(x + z_n) \ , \tag{3.8}$$

where we recall that we are still assuming (by contradiction) that $t_n = \|u_n\| \to \infty$. Therefore, since $\|\widetilde{v}_n\| = \|v_n\|$ is bounded, passing to a subsequence (if necessary), we have that

$$\begin{cases} \widetilde{v}_n \rightharpoonup \widetilde{v} \text{ weakly in } H^1(\mathbb{R}^N) \\ \widetilde{v}_n \to \widetilde{v} \text{ strongly in } L^q_{loc}(\mathbb{R}^N) \text{ (for any } 2 \leq q < 2^*) \\ \widetilde{v}_n \to \widetilde{v} \text{ a.e. in } \mathbb{R}^N . \end{cases}$$

where $\widetilde{v} \neq 0$ in view of (3.7).

Next, we divide (3.4) by t_n^2 to get

$$\frac{1}{2} - \int \frac{H(u_n)}{t_n^2} \, dx = o(1) \ ,$$

or, by translation invariance,

$$\frac{1}{2} - \int \frac{H(\widetilde{u}_n)}{t_n^2} \, dx = o(1) \ ,$$

or still, in view of (3.8),

$$\int \frac{H(\widetilde{u}_n)}{\widetilde{u}_n^2} \widetilde{v}_n^2 \, dx = \frac{\delta c_*}{2} + o(1) \ . \tag{3.9}$$

However, since $\lim_{|s| \to \infty} H(s)/s^2 = +\infty$ by (h_2) and since $\widetilde{v} \neq 0$, we conclude by the Fatou lemma that

$$\lim_{n \to \infty} \int \frac{H(\widetilde{u}_n)}{\widetilde{u}_n^2} \widetilde{v}_n^2 \, dx = \lim_{n \to \infty} \int \frac{H(t_n \widetilde{v}_n / \sqrt{\delta c_*})}{t_n^2 \widetilde{v}_n^2 / \delta c_*} \widetilde{v}_n^2 \, dx = +\infty \ ,$$

which is in clear contradiction with (3.9).

Finally, since either case $\alpha = 0$ or $\alpha > 0$ leads to a contradiction, we conclude that (u_n) is bounded in $H^1(\mathbb{R}^N)$. □

Our next result yields a bounded Palais–Smale sequence for φ at the level c_*.

Lemma 3.4. *Assume (h_1). Given any minimizing sequence (u_n) for c_*, there exists another minimizing sequence $(\widetilde{v}_n)$ such that $\|\nabla\varphi(\widetilde{v}_n)\| = o(1)$.*

Proof: Let (u_n) be a minimizing sequence for c_*. Then, (u_n) is bounded by Lemma 3.3 and, in view of the Ekeland variational principle [37] (see Appendix and Corollary A.3), there exists another minimizing sequence (v_n) for c_* such that

$$\|v_n - u_n\| = o(1) \tag{3.10}$$

and

$$\|[\nabla(\varphi|M)](v_n)\| = \|P_{T_{v_n}M}\nabla\varphi(v_n)\| = o(1) , \tag{3.11}$$

where P_{T_vM} denotes the orthogonal projection onto the tangent space to M at $v \in M$, that is, $P_{T_vM}h := h - \langle h, N_v\rangle N_v$, where $h \in H^1(\mathbb{R}^N)$ and $N_v := \frac{\nabla\psi(v)}{\|\nabla\psi(v)\|}$ is a unit normal to M at v. Clearly (3.10) implies that (v_n) is also a bounded sequence. The rest of the proof is somewhat technical and consists in showing that

$$\|\nabla\varphi(\widetilde{v}_n)\| = o(1)$$

for some other bounded minimizing sequence $(\widetilde{v}_n)$, as required. We start with the following

Claim 1: $\|\nabla\varphi(v_n)\|$ and $\|\nabla\psi(v_n)\|$ are bounded.

Indeed, let $2^\dagger = 2N/(N+2)$ be the conjugate exponent of 2^* and let $A = (-\Delta + I)^{-1} : L^{2^\dagger} \subset H^{-1} \longrightarrow H^1 \subset L^{2^*}$ be the bounded linear operator defined by $\langle Af, z\rangle := (f, z)_2 \;\; \forall z \in H^1$. Then, we have

$$\nabla\varphi(u) = u - A(h(u)) ,$$

$$\nabla\psi(u) = 2u - A(h(u) + h'(u)u) ,$$

so that (h_1) yields

$$\|A(h(u))\| \le ||h(u)||_{L^{2^\dagger}} \le ||u||_{L^{2^\dagger}} + C_1|||u|^{p-1}||_{L^{2^\dagger}} \le C(\|u\| + \|u\|^{p-1}) .$$

Similarly, we have

$$\|A(h'(u)u)\| \le C(\|u\| + \|u\|^{p-1}) .$$

Therefore, Claim 1 follows since $\|v_n\|$ is bounded.

Next, taking the inner-product of $P_{T_{v_n}M}\nabla\varphi(v_n)$ with $\nabla\varphi(v_n)$ and keeping in mind that $\|\nabla\varphi(v_n)\|$ is bounded, we obtain

$$\left\langle \nabla\varphi(v_n) - \langle \nabla\varphi(v_n), \frac{\nabla\psi(v_n)}{\|\nabla\psi(v_n)\|}\rangle \frac{\nabla\psi(v_n)}{\|\nabla\psi(v_n)\|}, \nabla\varphi(v_n) \right\rangle = o(1) ,$$

that is,

$$\|\nabla\varphi(v_n)\|^2 = \langle \nabla\varphi(v_n), \frac{\nabla\psi(v_n)}{\|\nabla\psi(v_n)\|}\rangle^2 + o(1) . \tag{3.12}$$

Similarly, taking the inner-product of $P_{T_{v_n}M}\nabla\varphi(v_n)$ with v_n and recalling that $\langle \nabla\varphi(v_n), v_n\rangle = \psi(v_n) = 0$ (since $v_n \in M$), we obtain

$$-\langle \nabla\varphi(v_n), \frac{\nabla\psi(v_n)}{\|\nabla\psi(v_n)\|}\rangle \langle \frac{\nabla\psi(v_n)}{\|\nabla\psi(v_n)\|}, v_n\rangle = o(1) ,$$

or

$$\frac{1}{\|\nabla\psi(v_n)\|} \cdot |\langle \nabla\varphi(v_n), \frac{\nabla\psi(v_n)}{\|\nabla\psi(v_n)\|}\rangle| \cdot |2\|v_n\|^2 - \int (h(v_n)v_n + h'(v_n)v_n^2)\, dx|$$

$$= A_n B_n C_n = o(1) , \tag{3.13}$$

where A_n is bounded from below away from zero by the previous claim.

Claim 2: $C_n > 0$ is bounded from below away from zero.

Indeed, since $\|v_n\|^2 = \int h(v_n)v_n\, dx$, we can write

$$C_n = \int h'(v_n)v_n^2\, dx - \int h(v_n)v_n\, dx .$$

Now, as in Lemma 3.3, we consider the quantity

$$\liminf_{n\to\infty} \left(\sup_{y\in\mathbb{R}^N} \int_{B_1(y)} |v_n|^2\, dx \right) := \alpha \tag{3.14}$$

and rule out the possibility $\alpha = 0$ as it would lead to the contradiction $\lim \int H(v_n)\, dx = \lim \int h(v_n)v_n\, dx = 0$ and $\lim \varphi(v_n) = 0$. Thus, we must have $\alpha > 0$ and

$$\int_{B_1(0)} |\widetilde{v}_n|^2\, dx \geq \frac{\alpha}{2} > 0 \tag{3.15}$$

for n large, where $\widetilde{v}_n(x) := v_n(x+z_n)$ for some $z_n \in \mathbb{R}^N$. As usual, by passing to a subsequence, if necessary, it follows that

$$\begin{cases} \widetilde{v}_n \rightharpoonup \widetilde{v} \text{ weakly in } H^1(\mathbb{R}^N) \\ \widetilde{v}_n \to \widetilde{v} \text{ strongly in } L^q_{loc}(\mathbb{R}^N) \text{ (for any } 2 \leq q < 2^*) \\ \widetilde{v}_n \to \widetilde{v} \text{ a.e. in } \mathbb{R}^N. \end{cases} \tag{3.16}$$

where $\widetilde{v} \neq 0$ in view of (3.15). Since we have

$$C_n = \int h'(\widetilde{v}_n)\widetilde{v}_n^2 \, dx - \int h(\widetilde{v}_n)\widetilde{v}_n \, dx$$

by translation invariance then, by assuming that $C_n = o(1)$, we would obtain by Remark 3.2 and the Fatou lemma that $\widetilde{v} = 0$, a contradiction. Thus, C_n is bounded from below away from zero and Claim 2 is proved.

Finally, since both A_n and C_n are bounded from below away from zero, we conclude from (3.13) that

$$B_n = |\langle \nabla\varphi(v_n), \frac{\nabla\psi(v_n)}{\|\nabla\psi(v_n)\|}\rangle| = o(1)$$

and, from (3.12) and translation invariance, that

$$\|\nabla\varphi(\widetilde{v}_n)\| = o(1) \, .$$

We have obtained a minimizing sequence $(\widetilde{v}_n) \subset M$ for c_* such that $\|\nabla\varphi(\widetilde{v}_n)\| = o(1)$, that is, a Palais–Smale sequence for φ at the level $c_* := \inf_{v \in M} \varphi(v)$. □

We are now ready for the main existence result of this section.

Theorem 3.5. *Assume* $(h_1), (h_2)$. *Then* $c_* = \inf_{v\in M} \varphi(v) > 0$ *is attained. In fact, there exists a minimizing sequence* $(\widetilde{v}_n)$ *that converges weakly to a minimizer of* c_*. *In particular, Problem* (3.1) *has a nonzero solution* $\widetilde{v} \in H^1(\mathbb{R}^N) \cap C^2(\mathbb{R}^N)$.

Proof: Let (u_n) be a minimizing sequence for c_*. In view of Lemma 3.3 and Lemma 3.4, there exists another minimizing sequence $(\widetilde{v}_n) \subset M$ satisfying (3.16). In particular, we have

$$\varphi(\widetilde{v}_n) = \frac{1}{2}\int [h(\widetilde{v}_n)\widetilde{v}_n - 2H(\widetilde{v}_n)] \, dx = c_* + o(1)$$

hence

$$\frac{1}{2}\int [h(\widetilde{v})\widetilde{v} - 2H(\widetilde{v})] \, dx \leq c_* \tag{3.17}$$

by convexity of $[sh(s) - 2H(s)]$ (cf. (h_2)). In addition, since

$$\varphi'(\widetilde{v}_n) \cdot \theta = \langle \widetilde{v}_n, \theta\rangle - \int h(\widetilde{v}_n)\theta \, dx = o(1)\|\theta\|$$

for all $\theta \in C_c^1(\mathbb{R}^N)$, and keeping (h_1) and (3.16) in mind, we obtain from the Vainberg theorem [72], Chapter 2.2, that $h(\tilde{v}_n) \longrightarrow h(\widetilde{v})$ in $L_{loc}^{\frac{q}{p-1}}(\mathbb{R}^N)$ for $p-1 \leq q < 2^*$ and, hence, that

$$\varphi'(\widetilde{v}) \cdot \theta = 0 \ ,$$

for all $\theta \in C_c^1(\mathbb{R}^N)$. Therefore, we conclude that $\widetilde{v} \neq 0$ is a (weak, hence strong) solution of (3.1), and it follows that $\widetilde{v} \in M$. In particular, by definition of c_* and from (3.17) we also conclude that $\varphi(\widetilde{v}) = c_*$. □

11

Lack of Compactness for Bounded Ω

In this chapter we present a simple situation of a variational problem on a bounded domain $\Omega \subset \mathbb{R}^N$ for which the corresponding functional does not satisfy the Palais–Smale condition $(PS)_c$ at certain levels $c \in \mathbb{R}$. We are referring to the so-called *strongly resonant problems*, a terminology introduced in Bartolo–Benci–Fortunato [8] for a resonant problem where the nonlinearity $g(s)$ was such that $\lim_{|s|\to\infty} g(s) = 0$ and $\lim_{|s|\to\infty} G(s) = \beta < \infty$. Such problems did not fall under the Landesman–Lazer [49] or the Ahmad–Lazer–Paul [3] framework. In [8], the authors used the weaker version of (PS) due to Cerami [24] in order to handle those problems (see condition (Cc) introduced in Exercise 2 at the end of Chapter 3). Here, we introduce a large class of *strongly resonant problems* and show that the standard (PS) condition can be used to handle such problems by determining the exact levels $c \in \mathbb{R}$ where it fails. We follow [7].

1 $(PS)_c$ for Strongly Resonant Problems

Let us consider the Dirichlet problem

$$\begin{cases} -\Delta u = \lambda_k u + g(x,u) & \text{in } \Omega \\ u = 0 \quad \text{on } \partial\Omega\,, \end{cases} \tag{1.1}$$

where Ω is a bounded domain in $\mathbb{R}^N$, $N \geq 3$, λ_k is an eigenvalue of $-\Delta$ in Ω with a Dirichlet boundary condition and $g : \Omega \times \mathbb{R} \longrightarrow \mathbb{R}$ is a Carathéodory function with subcritical growth. We call (1.1) a *strongly resonant problem* (at λ_k) if the following conditions hold:

$|g(x,s)| \leq h(x)$ for some $h \in L^q(\Omega)$, $q \geq \frac{2N}{N+2}$, and $g(x,s) \longrightarrow 0$ as $|s| \to \infty$, a.e. in Ω; $\qquad (g_1)$

$|G(x,s)| \leq h_1(x)$ for some $h_1 \in L^1(\Omega)$, and $G(x,s) \longrightarrow G_\pm(x)$ as $s \to \pm\infty$, a.e. in Ω. (g_2)

We remark that $(g_1), (g_2)$ hold if, for example, $\lim_{|s|\to\infty} g(x,s) = 0$ and $\lim_{|s|\to\infty} G(x,s) = G_\infty \in (-\infty, +\infty)$, uniformly for $x \in \Omega$. Such a situation is in sharp contrast with the one considered by Ahmad–Lazer–Paul (see Section 3 of Chapter 5), in which one had $G_\pm = +\infty$ (or $G_\pm = -\infty$). Indeed, the following lemma characterizes the values $c \in \mathbb{R}$ at which the functional

$$\begin{aligned} \varphi(u) &= \frac{1}{2}\int_\Omega (|\nabla u|^2 - \lambda_k u^2)\,dx - \int_\Omega G(x,u)\,dx \\ &= q(u) - N(u) \;, \quad u \in H_0^1(\Omega) \end{aligned} \tag{1.2}$$

corresponding to (1.1) satisfies $(PS)_c$. Moreover, it explains why $(PS)_c$ holds true for all $c \in \mathbb{R}$ in the Ahmad–Lazer–Paul situation.

Lemma 1.1. *Assume* $(g_1), (g_2)$. *Then,* φ *satisfies* $(PS)_c$ *if and only if*

$$c \notin \Gamma_k := \left\{ -\int_{[v>0]} G_+(x)\,dx - \int_{[v<0]} G_-(x)\,dx \mid v \in N_k \;,\; \|v\| = 1 \right\} ,$$

where $\|v\|^2 = \int_\Omega |\nabla v|^2\,dx$, *as usual, and* $N_k = N(L - \lambda_k)$ *denotes the eigenspace associated with* λ_k.

Proof: Let us denote by X^+, X^- the subspaces of $X := H_0^1(\Omega)$ where the quadratic form q is positive definite, negative definite, respectively, and write $X^0 = N_k$, so that

$$H_0^1(\Omega) = X^+ \oplus X^- \oplus X^0 \;.$$

Since g has subcritical growth, the functional φ satisfies $(PS)_c$ if and only if any sequence $\{u_n\} \subset H_0^1$ such that

(*i*) $\varphi(u_n) \longrightarrow c$,
(*ii*) $\varphi'(u_n) \longrightarrow 0$,

has a bounded subsequence. Let us assume that $\{u_n\}$ satisfies (*i*), (*ii*), but

(*iii*) $\|u_n\| \longrightarrow \infty$,

and prove that $c \in \Gamma_k$. Writing $u_n = u_n^+ + u_n^- + u_n^0$, where $u_n^+ \in X^+$, $u_n^- \in X^-$, $u_n^0 \in X^0 = N_k$, and using (*ii*), we have[1]

[1] We note that, in this chapter, the notations u^+ and u^- do not stand for the positive and negative parts of u.

$$|\langle\nabla\varphi(u_n),u_n^+\rangle| = |\,\|u_n^+\|^2 - \lambda_k\|u_n^+\|_{L^2}^2 - \int_\Omega g(x,u_n)u_n^+\,dx| \le \epsilon_n\|u_n^+\| \quad (1.3)$$

where $\epsilon_n = \|\varphi'(u_n)\|_{H^{-1}} \longrightarrow 0$. Using (g_1) and the fact that $q' \le 2N/(N-2)$, if $N \ge 3$ (where $q' = q/(q-1)$ denotes the *conjugate exponent* of q), we can estimate the integral term above as

$$|\int_\Omega g(x,u_n)u_n^+\,dx| \le \|h\|_{L^q}\|u_n^+\|_{L^{q'}} \le C\|h\|_{L^q}\|u_n^+\|, \quad (1.4)$$

in view of the Hölder and Sobolev inequalities. Therefore, (1.3) gives

$$(1-\frac{\lambda_k}{\lambda_{k+1}})\|u_n^+\|^2 \le (C\|h\|_{L^q}+\epsilon_n)\|u_n^+\| ,$$

so that $\{u_n^+\}$ is bounded in H_0^1. Similarly, we show that $\{u_n^-\}$ is also bounded. Thus, by (iii), we must have $\|u_n^0\| \to \infty$ and, letting $\widehat{u}_n := u_n/\|u_n^0\|$, it follows that $\widehat{u}_n \longrightarrow v \in N_k$ with $\|v\| = 1$. Without loss of generality, we may assume that $\widehat{u}_n(x) \longrightarrow v(x)$ a.e. in Ω. Therefore,

$$u_n(x) \longrightarrow +\infty \;\text{ a.e. in }\; [v>0] := \{x\in\Omega \mid v(x)>0\} ,$$

$$u_n(x) \longrightarrow -\infty \;\text{ a.e. in }\; [v<0] := \{x\in\Omega \mid v(x)<0\} .$$

Next, by (g_2), we can apply the Lebesgue theorem to the sequence $G(x,u_n(x))$ to obtain

$$\lim_{n\to\infty}\int_\Omega G(x,u_n(x))\,dx = \int_{[v>0]} G_+(x)\,dx + \int_{[v<0]} G_-(x)\,dx .$$

Therefore, in order to show that $c\in\Gamma_k$, it suffices to verify that $\|u_n^\pm\| \longrightarrow 0$. For that, we prove that the integral term in (1.3) (as well as the corresponding integral term in $|\langle\nabla\varphi(u_n),u_n^-\rangle|$) goes to zero.

Indeed, using the Hölder inequality and Sobolev embedding as in (1.4), we obtain (with $q \ge 2N/(N+2)$ if $N \ge 3$)

$$|\int_\Omega g(x,u_n)u_n^+\,dx| \le C(\int_\Omega |g(x,u_n(x))|^q\,dx)^{\frac{1}{q}}\|u_n^+\|$$

and, since $g(x,u_n(x)) \longrightarrow 0$ a.e in Ω with $|g(x,u_n(x))| \le h(x)$ a.e in Ω in view of (g_1), an application of the Lebesgue theorem once more implies the desired result that $c\in\Gamma_k$ if (i)–(iii) hold.

On the other hand, we also observe that φ does not satisfy $(PS)_c$ if $c\in\Gamma_k$. Indeed, taking $v\in N_k$, $\|v\| = 1$, we have

$$\varphi(tv) = -\int_\Omega G(x, tv(x))\, dx ,$$

$$\langle \nabla\varphi(tv), h\rangle = -\int_\Omega g(x, tv(x))h(x)\, dx ,$$

for any $t \in \mathbb{R}$ and $h \in H_0^1$. Letting $t \to +\infty$ and using (g_1), (g_2) together with the same arguments above, we conclude that

$$\varphi(tv) \longrightarrow -\int_{[v>0]} G_+(x)\, dx - \int_{[v<0]} G_-(x)\, dx ,$$

$$\varphi'(tv) \longrightarrow 0 . \qquad \square$$

Remark 1.1. Note that in the special case when $G_\pm(x) = G_\pm$ (constants) and λ_k is a simple eigenvalue, say $N_k = \text{span}\{\varphi_k\}$ with $\|\varphi_k\| = 1$, the set Γ_k has only two values (or perhaps one, if they coincide):

$$\Gamma_k = \{\ -G_+\alpha_k - G_-\beta_k\ ,\ -G_+\beta_k - G_-\alpha_k\ \} ,$$

where $\alpha_k = |[\varphi_k > 0]|$, $\beta_k = |[\varphi_k < 0]|$. Moreover, if $k = 1$, we get

$$\Gamma_1 = \{\ -G_+|\Omega|\ ,\ -G_-|\Omega|\ \} ,$$

where we are denoting the measure of a set S by $|S|$.

2 A Class of Indefinite Problems

In preparation for our existence result in the next section concerning a class of *strongly resonant* problems, we present in this section some preliminary results involving an elliptic problem with an indefinite linear part and an *almost monotone* nonlinearity. More precisely, we consider our previous Problem (1.1) in a bounded domain $\Omega \subset \mathbb{R}^N$ where the nonlinearity $g(x, u)$ is a *Carathéodory subcritical* function satisfying the lower estimate

$$(g(x, s) - g(x, t))(s - t) \geq -\delta(s - t)^2, \text{ a.e. } x \in \Omega, \forall s, t \in \mathbb{R}, \qquad (g_0)$$

for some $0 < \delta < \lambda_k - \lambda_{k-1}$, $k \geq 2$ (if $k = 1$ we only assume that $\lambda_1 > \delta > 0$). Here we are numbering the eigenvalues of $(-\Delta, H_0^1(\Omega))$ allowing repetition, but the above consecutive indices are taken such that $\lambda_{k-1} < \lambda_k$.

Remark 2.1. Note that (g_0) simply means that the right-hand side $f(x, u) = \lambda_k u + g(x, u)$ of (1.1) is a *monotone* function which *does not interact* with the eigenvalues $\lambda_1, \dots \lambda_{k-1}$. Therefore, in the present situation, it is natural to use the so-called *Liapunov–Schmidt decomposition method* to study our problem. As we shall see, this will reduce our pertinent indefinite functional

$$\varphi(u) = \frac{1}{2}\int_\Omega \left[(|\nabla u|^2 - \lambda_k u^2) - G(x,u)\right] dx = \frac{1}{2}\langle Lu, u\rangle - \int_\Omega G(x,u)\, dx \tag{2.1}$$

on $X := H_0^1(\Omega)$ to another functional $\psi(v)$ on a closed subspace V of X whose critical points are in a one-to-one correspondence with the critical points of $\varphi(u)$. In addition, for the class of strongly resonant problems to be considered in the next section, the reduced functional $\psi(v)$ will turn out to be bounded from below.

For our preliminary results, let us consider the previous orthogonal decomposition $X = X^- \oplus X^0 \oplus X^+$, where $X^0 = N_k = N(L - \lambda_k)$ is the kernel of the operator $L : H_0^1(\Omega) \longrightarrow H_0^1(\Omega)$ defined in (2.1), and X^+ (resp. X^-) denotes the subspace where L is positive definite (resp. negative definite). We denote the respective orthogonal projections by P_0, P_+ and P_-. Finally we will let

$$V = X^0 \oplus X^+ , \qquad W = X^- ,$$

so that $X = V \oplus W$, and we will write $P_V = P_0 + P_+$, $P_W = P_-$ for the corresponding orthogonal projections.

Lemma 2.1. *Assume* (g_0). *Then, for each* $v = u_0 + u_+ \in V$, *the functional* $W \ni w \longmapsto -\varphi(w+v)$ *is* μ*-convex on* W, *that is, the mapping*

$$w \longmapsto -P_W \nabla\varphi(w+v)$$

is strongly μ*-monotone on* W, *where* $\mu = \frac{\lambda_k - \delta}{\lambda_{k-1}} - 1 > 0$.

Proof: We want to show that

$$-\langle \nabla\varphi(w_1+v) - \nabla\varphi(w_2+v), w_1 - w_2\rangle \geq \mu\|w_1 - w_2\|^2 \quad \forall w_1, w_2 \in W,\ \forall v \in V .$$

Indeed, in view of (2.1) and (g_0), we have

$$\langle \nabla\varphi(w_1+v) - \nabla\varphi(w_2+v), w_1 - w_2\rangle$$

$$= \langle L(w_1+v) - L(w_2+v), w_1 - w_2\rangle - \int_\Omega [g(x, w_1+v) - g(x, w_2+v)](w_1 - w_2)\, dx$$

$$= \int_\Omega [(\nabla(w_1+v) - \nabla(w_2+v))\cdot(\nabla(w_1 - w_2)) - \lambda_k((w_1+v) - (w_2+v))(w_1 - w_2)]\, dx$$

$$
\begin{aligned}
&- \int_\Omega [g(x, w_1 + v) - g(x, w_2 + v)](w_1 - w_2)\,dx \\
&\le \|w_1 - w_2\|^2 - \lambda_k \int_\Omega |w_1 - w_2|^2\,dx + \delta \int_\Omega |w_1 - w_2|^2\,dx \\
&= \|w_1 - w_2\|^2 - (\lambda_k - \delta) \int_\Omega |w_1 - w_2|^2\,dx \\
&\le (1 - \frac{\lambda_k - \delta}{\lambda_{k-1}}) \|w_1 - w_2\|^2 = -\mu \|w_1 - w_2\|^2
\end{aligned}
$$

where, in the last inequality, we used the fact that $\int_\Omega |\nabla w|^2 \le \lambda_{k-1} \int_\Omega |w|^2$ for all $w \in W = X_-$. □

Now, in view of Lemma 2.1 and a classical result for μ-monotone maps (see [4, 22]), for each $v \in V$ there exists a unique $w = \theta(v) \in W$ such that

$$P_W \nabla \varphi(\theta(v) + v) = 0 \;.$$

Moreover, it is shown in [4, 22] that the map $V \ni v \longmapsto \theta(v) \in W$ is continuous, the functional

$$\psi(v) = \varphi(\theta(v) + v) = \sup_{w \in W} \varphi(w + v)$$

is of class C^1 (together with φ), and

$$\nabla \varphi(u) = 0 \quad \text{if and only if} \quad u = \theta(v) + v \quad \text{and} \quad \nabla \psi(v) = 0 \;. \tag{2.2}$$

Remark 2.2. Unlike in the usual Liapunov–Schmidt reductions, note that our reduced functional ψ is defined on the *infinite-dimensional* space V.

Lemma 2.2. *The reduced functional* $\psi : V \longrightarrow \mathbb{R}$ *is weakly* l.s.c..

Proof: It suffices to observe that

$$\psi(\cdot) = \sup_{w \in W} \varphi(w + \cdot)$$

is the supremum of the family $\{\varphi(w+\cdot) \mid w \in W\}$ and each functional $\varphi(w+\cdot)$ is weakly lower-semicontinuous on V. □

Lemma 2.3. *If* $\varphi : X \longrightarrow \mathbb{R}$ *satisfies* $(PS)_c$ *for some* $c \in \mathbb{R}$*, then also* $\psi : V \longrightarrow \mathbb{R}$ *satisfies* $(PS)_c$.

Proof: Let $v_n \in V$ be such that

$$\psi(v_n) \longrightarrow c \quad , \quad \nabla\psi(v_n) \longrightarrow 0 \ .$$

Letting $w_n := \theta(v_n)$, this means that

$$\varphi(w_n + v_n) \longrightarrow c \quad , \quad P_V \nabla\varphi(w_n + v_n) \longrightarrow 0 \ .$$

Since by definition $P_W \nabla\varphi(w_n + v_n) = 0$, it follows that

$$\varphi(w_n + v_n) \longrightarrow c \quad , \quad \nabla\varphi(w_n + v_n) \longrightarrow 0 \ ,$$

and hence, up to a subsequence, we have that

$$(w_n, v_n) \longrightarrow (w_\infty, v_\infty)$$

for some $w_\infty \in W$, $v_\infty \in V$. □

3 An Application

Still considering our original Problem (1.1) with the nonlinearity g being *almost monotone* (i.e., satisfying condition (g_0)), we observe that the functional

$$u \longmapsto -N(u) = -\int_\Omega G(x, u)\, dx \tag{3.1}$$

is bounded from below whenever $\limsup_{s\to\pm\infty} G(x.s) < +\infty$ uniformly a.e. $x \in \Omega$. In fact, we are going to assume that $g : \overline{\Omega} \times \mathbb{R} \longrightarrow \mathbb{R}$ is continuous and satisfies

$$\lim_{|s|\to\infty} g(x, s) = 0 \quad \text{uniformly for a.e. } x \in \Omega \ , \tag{g_1}$$
$$\lim_{|s|\to\infty} G(x, s) = 0 \quad \text{uniformly for a.e. } x \in \Omega \ , \tag{g_2}$$

so that Problem (1.1) is a *strongly resonant problem* at the eigenvalue λ_k.

Lemma 3.1. *Assume g satisfies (g_0) (i.e., g is almost monotone).*

(*i*) *If $g(x, 0) = 0$, then $\psi(0) = 0$;*
(*ii*) *If (g_2) holds, then $\psi(u^0) \longrightarrow 0$ as $\|u^0\| \to \infty$, $u^0 \in X^0 = N_k$.*

Proof: (*i*) Since $u = 0$ is a solution of Problem (1.1), we have by (2.2) that $\nabla\varphi(0) = 0$, hence $v = 0$, $\theta(0) = 0$. In particular, $\psi(0) = 0$.

(*ii*) Let us recall the definition of $N(u)$ and the quadratic form $q(u)$ in (1.2). Then

$$\psi(u^0) = \varphi(\theta(u^0) + u^0) = q(\theta(u^0)) - N(\theta(u^0) + u^0) \tag{3.2}$$

since $q(u^0) = 0$. On the other hand, we have

$$\psi(u^0) = \sup_{w \in W} \varphi(w + u^0) \geq \varphi(u^0).$$

Therefore, it follows from (3.2) that

$$|q(\theta(u^0))| = -q(\theta(u^0)) \leq -\varphi(u^0) - N(\theta(u^0) + u^0) = N(u^0) - N(\theta(u^0) + u^0) ,$$

hence

$$|q(\theta(u^0))| \leq N(u^0) - N(\theta(u^0) + u^0) , \tag{3.3}$$

which shows that $|q(\theta(u^0))|$ is bounded in view of (g_2). Since $-q$ is coercive on W, we conclude that

$$\|\theta(u^0)\| \leq C . \tag{3.4}$$

Next, we prove the following

Claim: $\lim_{\|u^0\| \to \infty} N(\theta(u^0) + u^0) = \lim_{\|u^0\| \to \infty} N(u^0) = 0$.
Indeed, assuming $\|u_n^0\| \longrightarrow \infty$, we have in view of (3.4) that

$$\|\theta(u_n^0) + u_n^0\| \geq \|u_n^0\| - \|\theta(u_n^0)\| \geq \|u_n^0\| - C ,$$

hence

$$\|\theta(u_n^0) + u_n^0\| \longrightarrow \infty .$$

Letting $v_n := (\theta(u_n^0) + u_n^0)/\|\theta(u_n^0) + u_n^0\| \in W \oplus N(L - \lambda_k)$ and recalling that $W \oplus N(L - \lambda_k)$ is finite dimensional, we may assume (passing to a subsequence, if necessary) that $v_n \longrightarrow v$ for some $v \in W \oplus N(L - \lambda_k)$ with $\|v\| = 1$. Thus $v(x) \not\equiv 0$ a.e. and $\theta(u_n^0) + u_n^0 \longrightarrow +\infty$ on $[v > 0]$ (resp. $\theta(u_n^0) + u_n^0 \longrightarrow -\infty$ on $[v < 0]$). In view of (g_2), an application of Lebesgue's theorem gives $\lim_{\|u^0\| \to \infty} N(\theta(u^0) + u^0) = 0$. Similarly, we show that $\lim_{\|u^0\| \to \infty} N(u^0) = 0$, which proves the claim.

Finally, from (3.3), (3.2), and the above claim we conclude that

$$\lim_{\|u^0\| \to \infty} \psi(u^0) = 0 .$$

□

Now, for each $r > 0$, let us denote by C_r the cylinder in V given by

$$C_r := \{ \, v = u^0 + u^+ \in V \mid u^0 \in X^0 \, , \, \|u^0\| = r \, , \, u^+ \in X^+ \, \}$$

and define

$$m_r := \inf_{v \in C_r} \psi(v) \ .$$

We note that

$$\psi(u^0 + u^+) \geq \varphi(u^0 + u^+) = q(u^+) - N(u^0 + u^+) \ ,$$

where the above quadratic term is coercive on X^+ and the latter nonlinear term is uniformly bounded in view of assumption (g_2). Therefore, ψ is coercive on each C_r and, since ψ is weakly lower-semicontinuous by Lemma 2.2 , it follows that

$$m_r := \inf_{v \in C_r} \psi(v) > -\infty$$

and, in fact, m_r is attained. Also note that

$$m := \inf_{v \in V} \psi(v) > -\infty$$

and, clearly,

$$m \leq m_r \quad \text{for all } r > 0. \tag{3.5}$$

We are now ready to prove the main existence result of this section.

Theorem 3.2. *Let $g(x,0) = 0$, so that $u = 0$ is a solution of* (1.1). *If* (g_0), (g_1), *and* (g_2) *hold then* (1.1) *has a nonzero solution $u \in H_0^1$.*

Proof: In view of Lemma 1.1 the functional φ satisfies $(PS)_c$ for all $c \neq 0$. It follows from Lemma 2.3 that the reduced functional $\psi : V \longrightarrow \mathbb{R}$ also satisfies $(PS)_c$ for all $c \neq 0$ and, from Lemma 3.1(i), that

$$m \leq \psi(0) = 0 \ .$$

Case 1: $m < 0$. In this case, since ψ satisfies $(PS)_m$, we have that $m < 0 = \psi(0)$ is a critical value of ψ, that is, there exists $\widehat{v} \in V$ such that

$$\psi(\widehat{v}) = m < 0 \quad , \quad \nabla\psi(\widehat{v}) = 0 \ .$$

In particular, $\widehat{v} \neq 0$ since $\psi(0) = 0$ by Lemma 3.1(i). It follows from (2.2) that

$$\nabla\varphi(\theta(\widehat{v}) + \widehat{v}) = 0 \ ,$$

that is, $\widehat{u} = \theta(\widehat{v}) + \widehat{v} \neq 0$ is a solution of (1.1).

Case 2: $m = 0$. In this case fix $R > 0$ and note that, by (3.5), we have

$$m_R \geq 0 \ .$$

If $m_R = 0$, then $\psi(\widehat{v}_R) = 0$ for some $\widehat{v}_R = u_R^0 + u_R^+ \in C_R$. Since we are assuming in this case that $m = \inf_{v \in V} \psi(v) = 0$, we conclude that $\widehat{v}_R \neq 0$ is a critical point of ψ and, as in Case 1, that $\widehat{u}_R = \theta(\widehat{v}_R) + \widehat{v}_R$ is a nonzero solution of (1.1).

On the other hand, let us assume that $m_R > 0$. Then, by Lemma 3.1(ii), there exists $\widetilde{u}^0 \in N_k$ such that $\|\widetilde{u}^0\| > R$ and

$$\psi(\widetilde{u}^0) < m_R \ .$$

Consider the class of all paths in V joining 0 and $\widetilde{u}^0$, i.e.,

$$\widetilde{\Gamma} = \{ \ \gamma \in C([0,1], V) \mid \gamma(0) = 0 \ , \ \gamma(1) = \widetilde{u}^0 \ \} \ ,$$

and define

$$\widetilde{c} := \inf_{\gamma \in \widetilde{\Gamma}} \max_{t \in [0,1]} \psi(\gamma(t)) \ .$$

Clearly $\gamma([0,1]) \cap C_R \neq \emptyset$ for any $\gamma \in \widetilde{\Gamma}$ and, hence,

$$\widetilde{c} \geq m_R > 0$$

Since ψ satisfies $(PS)_{\widetilde{c}}$, it follows as in the proof of the mountain-pass theorem that $\widetilde{c} > 0$ is a critical value of ψ, that is,

$$\psi(\widetilde{v}) = m_R > 0 \quad , \quad \nabla\psi(\widetilde{v}) = 0$$

for some $0 \neq \widetilde{v} \in V$. Therefore, as in Case 1, we conclude that $\widetilde{u} = \theta(\widetilde{v}) + \widetilde{v}$ is a nonzero solution of (1.1). □

Remark 3.1. We point out that the above proof shows more than existence of a nonzero solution $u \in H_0^1$ for problem (1.1). Namely, it shows that

(i) either (1.1) has a solution $u \in H_0^1$ at an *energy level* $c = \varphi(u) \neq 0$,
(ii) or else, (1.1) has uncountably many solutions $u_r \in H_0^1$, $r > 0$, at the *zero energy level* $c = \varphi(u_r) = 0$.

12

Appendix

1 Ekeland Variational Principle

In this appendix we present a beautiful result, due to I. Ekeland, that gives the most information one can get from a *lower-semicontinuous* function $\varphi : M \longrightarrow \mathbb{R} \cup \{+\infty\}$ which is *bounded from below* on a *complete metric space.* It is his celebrated variational principle [37], which has far reaching applications in various areas of analysis and, among other things, explains the significance of the *Palais–Smale condition.*

Ekeland Variational Principle. *Let* (M, d) *be a complete metric space and* $\varphi : M \longrightarrow \mathbb{R} \cup \{+\infty\}$ *be a lower-semicontinuous function which is bounded from below. Suppose* $\epsilon > 0$ *and* $\hat{u}$ *are such that*

$$\varphi(\hat{u}) \leq \inf_M \varphi + \epsilon. \tag{1.1}$$

Then, given any $\lambda > 0$, *there exists* $u_\lambda \in M$ *such that*

$$\varphi(u_\lambda) \leq \varphi(\hat{u}), \tag{1.2}$$

$$d(u_\lambda, \hat{u}) \leq \lambda, \tag{1.3}$$

and

$$\varphi(u_\lambda) < \varphi(u) + \frac{\epsilon}{\lambda} d(u_\lambda, u) \qquad \forall\, u \neq u_\lambda\,. \tag{1.4}$$

Proof: Writing $d_\lambda = \frac{d}{\lambda}$ for simplicity, it is easy to see that

$$u \leq v \quad \text{if and only if} \quad \varphi(u) \leq \varphi(v) - \epsilon d_\lambda(u,v)$$

defines a partial order on M. Then, letting $u_1 = \hat{u}$, let us define recursively a nonincreasing sequence of subsets of M,

$$S_1 \supset S_2 \supset \cdots \supset S_n \supset \cdots,$$

by setting

$$S_1 := \{u \in M \mid u \leq u_1\}$$

$$S_2 := \{u \in M \mid u \leq u_2\} \text{ where } u_2 \in S_1 \text{ satisfies } \varphi(u_2) \leq \inf_{S_1} \varphi + \frac{\epsilon}{2},$$

and so on,

$$S_n := \{u \in M \mid u \leq u_n\} \text{ where } u_n \in S_{n-1} \text{ satisfies } \varphi(u_n) \leq \inf_{S_{n-1}} \varphi + \frac{\epsilon}{2^{n-1}}.$$

Claim: (i) Each S_n is *closed*; (ii) $\operatorname{diam}(S_n) \to 0$ as $n \to \infty$. Indeed:

(i) If $w_k \to w \in M$ and $w_k \in S_n$, then $w_k \leq u_n$ and

$$\varphi(w) \leq \liminf_{k\to\infty} \varphi(w_k) \leq \varphi(u_n) - \epsilon \lim_{k\to\infty} d_\lambda(w_k, u_n) = \varphi(u_n) - \epsilon d_\lambda(w, u_n)$$

by the lower-semicontinuity of φ, hence $w \in S_n$.

(ii) If $w \in S_n$, then $w \leq u_n$ and, by the choice of u_n, it follows that

$$\varphi(w) \leq \varphi(u_n) - \epsilon d_\lambda(w, u_n)$$

and

$$\varphi(u_n) \leq \varphi(w) + \frac{\epsilon}{2^{n-1}}.$$

These two last inequalities imply that

$$d_\lambda(w, u_n) \leq \frac{1}{2^{n-1}} \qquad \forall w \in S_n,$$

so that $\lim_{n\to\infty} \operatorname{diam}(S_n) = 0$. Since M is complete, one has

$$\bigcap_{n=1}^{\infty} S_n = \{u_\lambda\}$$

for a unique $u_\lambda \in M$. Now, we claim that u_λ satisfies (1.2), (1.3) and (1.4). Indeed, (1.2) is obvious and, by uniqueness of u_λ, one has $u \not\leq u_\lambda$ for all $u \neq u_\lambda$. In other words, one has

$$\varphi(u) > \varphi(u_\lambda) - \epsilon d_\lambda(u, u_\lambda) \quad \forall u \neq u_\lambda \,,$$

which shows that (1.4) holds true. Finally, (1.3) follows by passing to the limit in the inequality

$$d_\lambda(\hat{u}, u_n) \leq \sum_{j=1}^{n-1} d_\lambda(u_j, u_{j+1}) \leq \sum_{j=1}^{n-1} \frac{1}{2^j} \,. \qquad \square$$

A useful choice of λ occurs when $\lambda = \sqrt{\epsilon}$, namely, one has the following:

Corollary A.1 *Let (M, d) be a complete metric space and $\varphi : M \longrightarrow \mathbb{R} \cup \{+\infty\}$ be a lower-semicontinuous function which is bounded from below. Suppose $\epsilon > 0$ and $\hat{u}$ are such that*

$$\varphi(\hat{u}) \leq \inf_M \varphi + \epsilon.$$

Then, there exists $\hat{v} \in M$ such that

$$\varphi(\hat{v}) \leq \varphi(\hat{u}) \,,$$

$$d(\hat{v}, \hat{u}) \leq \sqrt{\epsilon} \,,$$

and

$$\varphi(\hat{v}) < \varphi(u) + \sqrt{\epsilon}\, d(\hat{v}, u) \quad \forall\, u \neq \hat{v} \,.$$

When M has a linear structure, say M is a Banach space X, and φ is differentiable, one obtains the following consequence of Ekeland variational principle:

Corollary A.2 *Let X be a Banach space and $\varphi : X \longrightarrow \mathbb{R}$ be a differentiable functional which is bounded from below. Given a minimizing sequence u_n for φ, there exists another minimizing sequence v_n such that*

$$\varphi(v_n) \leq \varphi(u_n) \,, \tag{1.5}$$

$$\|v_n - u_n\| \longrightarrow 0 \,, \tag{1.6}$$

and

$$\|\varphi'(v_n)\|_{X^*} \longrightarrow 0 \,. \tag{1.7}$$

Proof: By applying the previous corollary with $\epsilon_n = \max\{\frac{1}{n}, \varphi(u_n) - c\}$, where $c = \inf_X \varphi$, we obtain (1.5) and (1.6) readily, as well as the inequality

$$\varphi(v_n) < \varphi(u) + \sqrt{\epsilon_n}\,\|v_n - u\| \quad \forall\, u \neq v_n\,.$$

Now, the choice $u = v_n + th$ with $\|h\| = 1$ and $t > 0$ yields

$$\varphi(v_n + th) - \varphi(v_n) > -t\sqrt{\epsilon_n}\,,$$

so that, dividing by t and letting $t \to 0$ one gets

$$\varphi'(v_n) \cdot h \geq -\sqrt{\epsilon_n}$$

for all $h \in X$ with $\|h\| = 1$. This clearly implies (1.7). □

A particular situation which we used in Chapter 10 involved a C^1-functional on a Hilbert space E, and a closed C^1-submanifold $M \subset E$ of codimension 1 defined by a level set of another C^1-functional on E. Namely:

Corollary A.3 *Let $\varphi \in C^1(E, \mathbb{R})$ (E a Hilbert space) and let $M = \{u \in E \,|\, \psi(u) = 0\}$*[1]*, where $\psi \in C^1(E, \mathbb{R})$ with $\nabla\psi(u) \neq 0$ for all $u \in M$. Assume that φ is bounded from below on M, and let $u_n \in M$ be a minimizing sequence for $\varphi|M$. Then, there exists another minimizing sequence $v_n \in M$ such that*

$$\varphi(v_n) \leq \varphi(u_n)\,, \tag{1.8}$$

$$\|v_n - u_n\| \longrightarrow 0, \tag{1.9}$$

and

$$\|\nabla(\varphi|M)(v_n)\| \longrightarrow 0\,, \tag{1.10}$$

as $n \to \infty$.

Proof: As before, by applying Corollary A.1 with $\epsilon_n = \max\{\frac{1}{n}, \varphi(u_n) - c\}$, where now $c = \inf_M \varphi$, we readily obtain (1.8) and (1.9), as well as the inequality

$$\varphi(v_n) < \varphi(u) + \sqrt{\epsilon_n}\,\|v_n - u\| \quad \forall\, u \neq v_n\,. \tag{1.11}$$

Now, for each $v_n \in M$ and each tangent vector $h \in T_{v_n}M$, we have a mapping $\gamma : (-1, 1) \longrightarrow M$ such that $\gamma(0) = v_n$ and $\gamma'(0) = h$. From (1.11) with $u = \gamma(t)$, we obtain

[1] Actually, in Chapter 10, we removed an isolated point ($u = 0$) from M.

$$\varphi(\gamma(t)) - \varphi(v_n) > -\sqrt{\epsilon_n}\, \|\gamma(t) - v_n\|\,,$$

so that, dividing by $t > 0$ and letting $t \to 0$ one gets

$$\langle \nabla\varphi(v_n), \gamma'(0)\rangle \geq -\sqrt{\epsilon_n}\, \|\gamma'(0)\|\,.$$

Therefore, from this it follows that

$$\|\nabla(\varphi|M)(v_n)\| \leq \sqrt{\epsilon_n} \longrightarrow 0 \quad as\ \ n \to \infty\,,$$

and (1.10) also holds true. □

Remark 1.1. Given $v \in M$, we note that $\nabla(\varphi|M)(v)$ is the projection of $\nabla\varphi(v)$ over the tangent space $T_v M$, that is,

$$\nabla(\varphi|M)(v) = \nabla\varphi(v) - \langle \nabla\varphi(v), \frac{\nabla\psi(v)}{\|\nabla\psi(v)\|}\rangle \frac{\nabla\psi(v)}{\|\nabla\psi(v)\|}\,.$$

References

[1] R.A. Adams, *Sobolev Spaces*, Academic Press, 1976.
[2] S. Agmon, *The L^p approach to the Dirichlet problem*, Ann. Scuola Norm. Sup. Pisa 13 (1959), 405–448.
[3] S. Ahmad, A.C. Lazer and J.L. Paul, *Elementary critical point theory and perturbations of elliptic boundary value problems at resonance*, Indiana Univ. Math. J. 25 (1976), 933–944.
[4] H. Amann, *Saddle points and multiple solutions of differential equations*, Math. Z. 169 (1979), 127–166.
[5] H. Amann, *Ljusternik–Schnirelman theory and nonlinear eigenvalue problems*, Math. Ann. 199 (1972), 55–72.
[6] A. Ambrosetti, *On the existence of multiple solutions for a class of nonlinear boundary value problems*, Rend. Sem. Mat. Padova 49 (1973), 195–204.
[7] D. Arcoya and D.G. Costa, *Nontrivial solutions for a strong resonant problem*, Diff. Int. Eqs. 8 (1995), 151–159.
[8] P. Bartolo, V. Benci and D. Fortunato, *Abstract critical point theorems and applications to some nonlinear problems with strong resonance at infinity*, Nonl. Anal. 7 (1983), 981–1012.
[9] A. Ambrosetti and P.H. Rabinowitz, *Dual variational methods in critical point theory and applications*, J. Funct. Anal. 14 (1973), 349–381.
[10] V. Benci, *A geometrical index for the group S^1 and some applications to the study of periodic solutions of ordinary differential equations*, Comm. Pure Appl. Math. 34 (1981), 393–432.
[11] V. Benci, *On critical point theory for indefinite functionals in the presence of symmetries*, Trans. Amer. Math. Soc. 274 (1982), 533–572.
[12] V. Benci, *Some critical point theorems and applications*, Comm. Pure Appl. Math. 33 (1980), 147–172.
[13] V. Benci and P.H. Rabinowitz, *Critical point theorems for indefinite functionals*, Invent. Math. 52 (1979), 241–247.
[14] H. Berestycki and P.-L. Lions, *Nonlinear scalar field equations*, Arch. Rat. Mech. Anal. 82 (1983), 313–376.
[15] M.S. Berger, *Nonlinearity and Functional Analysis*, Academic Press, 1977.
[16] M.S. Berger, *Periodic solutions of second order dynamical systems and isoperimetric variational problems*, Amer. J. Math. 93 (1971), 1–10.

[17] H. Brézis, *Periodic solutions of nonlinear vibrating strings and duality principles*, Bull. Amer. Math. Soc. 8 (1983), 409–426.

[18] H. Brézis, J.M. Coron and L. Nirenberg, *Free vibrations for a nonlinear wave equation and a theorem of P. Rabinowitz*, Comm. Pure Appl. Math. 33 (1980), 667–689.

[19] H. Brézis and E. Lieb, *A relation between pointwise convergence of functions and convergence of functionals*, Proc. Amer. Math. Soc. 88 (1983), 486–490.

[20] H. Brézis and L. Nirenberg, *Remarks on finding critical points*, Comm. Pure Appl. Math. 44 (1991), 939–963.

[21] F.E. Browder, *Nonlinear eigenvalue problems and group invariance, in Functional Analysis and Related Fields* (F.E. Browder, ed.), Springer-Verlag, 1970, 1–58.

[22] A. Castro, *Metodos Variacionales y Analisis Funcional no Lineal*, X Coloquio Colombiano de Matematicas, 1980.

[23] A. Castro and A.C. Lazer, *Applications of a max-min principle*, Rev. Colombiana Mat. 10 (1976), 141–149.

[24] G. Cerami, *Un criterio de esistenza per i punti critici su varietà ilimitate*, Rx. Ist. Lomb. Sci. Lett. 112 (1978), 332–336.

[25] D.C. Clark, *A variant of the Lusternik–Schnirelman theory*, Ind. Univ. Math. J. 22 (1972), 65–74.

[26] F. Clarke and I. Ekeland, *Hamiltonian trajectories having prescribed minimal period*, Comm. Pure Appl. Math. 33 (1980), 103–166.

[27] C.V. Coffman, *A minimum-maximum principle for a class of nonlinear integral equations*, J. Analyse Math. 22 (1969), 391–419.

[28] J.M. Coron, *Résolution de l'équation $Au + Bu = f$, où A est linéaire autoadjoint et B est un opérateur potential non linéaire* C. R. Acad. Sc. Paris 288 (1979), 805–808.

[29] D.G. Costa, *Topicos em Análise Não-Linear e Aplicações às Equações Diferenciais*, VIII Escola Latino-Americana de Matemática, IMPA-CNPq, Rio de Janeiro, 1986.

[30] D.G. Costa and M. Willem, *Multiple critical points of invariant functionals and applications*, Nonl. Anal. 10 (1986), 843–852.

[31] D.G. Costa and C.A. Magalhaes, *Variational elliptic problems which are nonquadratic at infinity*, Nonl. Anal. 23 (1994), 1401–1412.

[32] D.G. Costa, Y. Guo and M. Ramos, *Existence and multiplicity results for indefinite elliptic problems in $\mathbb{R}^N$*, Elect. J. Diff. Eqns. 25 (2002), 1–15.

[33] R. Courant and D. Hilbert, *Methods of Mathematical Physics*, Vol. I, Wiley, 1962.

[34] R. Courant and D. Hilbert, *Methods of Mathematical Physics*, Vol. II, Wiley, 1962.

[35] A. Dolph, *Nonlinear integral equations of Hammerstein type*, Trans. Amer. Math. Soc. 66 (1949), 289–307.

[36] N. Dunford and J.T. Schwartz, *Linear Operators—Part I*, Interscience Publishers, Inc., 1957.

[37] I. Ekeland, *On the variational principle*, J. Math. Anal. Appl. 47 (1974), 324–353.

[38] I. Ekeland and J.-M. Lasry, *On the number of periodic trajectories for a Hamiltonian flow on a convex energy surface*, Ann. Math. 112 (1980), 283–319.

[39] I. Ekeland and J.-M. Lasry, *Advances in Hamiltonian Systems*, Birkhäuser, 1983, p. 74.

[40] I. Ekeland and R. Teman, *Convex Analysis and Variational Problems*, North Holland Publ., Amsterdam, 1976.

[41] E.R. Fadell and P.H. Rabinowitz, *Generalized cohomological index theories for Lie group actions with an application to bifurcation questions for Hamiltonian systems*, Invent. Math. 45 (1978), 139–174.

[42] E.R. Fadell, S.Y. Husseini and P.H. Rabinowitz, *Borsuk–Ulam theorems for arbitrary S^1 actions and applications*, Trans. Amer. Math. Soc. 274 (1982), 345–360.

[43] D.G. de Figueiredo, *Lectures on the Ekeland Variational Principle with Applications and Detours*, Springer-Verlag, 1989.

[44] D.G. de Figueiredo, *O Princípio de Dirichlet*, Matemática Universitária 1, Soc. Bras. Mat. 4 (1985), 68–84.

[45] E.W.C. van Groesen, *Applications of natural constraints in critical point theory to periodic solutions of natural Hamiltonian systems*, MRC Technical Report # 2593, 1983.

[46] A. Hammerstein, *Nichlineare Integralgleichungen nebst Anwendungen*, Acta Math. 54 (1930), 117–176.

[47] M.A. Krasnoselskii, *Topological Methods in the Theory of Nonlinear Integral Equations*, Pergamon Press, 1964.

[48] S. Lang, *Analysis II*, Addison-Wesley Publishing Company, Inc., 1969.

[49] E.M. Landesman and A.C. Lazer, *Nonlinear perturbations of linear elliptic boundary value problems at reonance*, J. Math. Mech. 19 (1970), 609–623.

[50] J. Leray and J. Schauder, *Topologie et équations fonctionelles*, Ann. Scuola Norm. Sup. Pisa 3 (1934), 45–78.

[51] P.-L. Lions, *The concentration-compactness principle in the calculus of variations. The locally compact case*, Ann. Inst. Henri Poincaré, Analyse Non Linéaire 1 (1984), 109–145 and 223–283.

[52] P.-L. Lions, *The concentration-compactness principle in the calculus of variations. The limit case*, Rev. Mat. Iberoamericana I (1985), 145–201, and 2 (1985), 45–121.

[53] L. Lusternik, *Topologische Grundlagen der allgemeinen Eigenwerttheorie*, Monatsch. Math. Phys. 37 (1930), 125–130

[54] L. Lusternik and L. Schnirelman, *Topological Methods in the Calculus of Variations*, Herman, 1934.

[55] J. Mawhin and M. Willem, *Critical Point Theory and Hamiltonian Systems*, Applied Mathematical Sciences 74, Springer-Verlag, New York, 1989.

[56] J. Mawhin, J.R. Ward Jr. and M. Willem, *Variational methods and semilinear elliptic equations*, Arch. Rat. Mech. Anal. 95 (1986), 269–277.

[57] L. Nirenberg, *Topics in Nonlinear Functional Analysis*, Courant Lecture Notes in Mathematics 6, Courant Institute of Mathematical Sciences, American Mathematical Society, New York, 2001 (revised reprint of 1974 original).

[58] L. Nirenberg, *Variational and topological methods in nonlinear problems*, Bull. Amer. Math. Soc. 4 (1981), 267–302.

[59] R.S. Palais, *Lusternik–Schnirelman theory on Banach manifolds*, Topology 5 (1966), 115–132.

[60] R.S. Palais, *Critical point theory and the minimax principle*, in Proc. Symp. Pure Math. XV, AMS, 1970, 1970, 185–212.

[61] R.S. Palais, *The principle of symmetric criticality*, Comm. Math. Phys. 69 (1979), 19–30.

[62] R.S. Palais and S. Smale, *A generalized Morse theory*, Bull. Amer. Math. Soc. 70 (1964), 165–171.

[63] P.H. Rabinowitz, *Minimax Methods in Critical Point Theory with Applications to Differential Equations*, CBMS Regional Conf. Ser. in Math. 65, AMS, Providence, RI, 1986.

[64] P.H. Rabinowitz, *Some minimax theorems and applications to nonlinear partial differential equations*, in Nonlinear Analysis: A Collection of Papers in Honor of Erich N. Rothe (L. Cesari, R. Kannan, H.F. Weinberger, editors), Academic Press (1978), 161–177.

[65] P.H. Rabinowitz, *The Mountain-Pass Theorem: Theme and Variations*, in Differential Equations (D.G. de Figueiredo, C.S. Honig, editors), Lecture Notes in Mathematics 957, Springer-Verlag, 1982.

[66] P.H. Rabinowitz, *Variational methods for nonlinear eigenvalue problems*, in Nonlinear Eigenvalue Problems (G. Prodi, editor) CIME, Edizioni Cremonese, Rome, 1974, 141–195.

[67] M. Schechter, *A variation of the mountain-pass lemma and applications*, J. London Math. Soc. (2) 44 (1991), 491–502.

[68] L. Schnirelman, *Uber eine neue kombinatorische Invariante*, Monatsch. Math. Phys. 37 (1930), no. 1.

[69] M. Struwe, *Variational Methods and Applications to Nonlinear Partial Differential Equations and Hamiltonian Systems*, Springer-Verlag, Berlin, 1990.

[70] F. Trèves, *Basic Linear Partial Differential Equations*, Academic Press, 1975.

[71] M.M. Vainberg, *Variational Methods in the Theory of Nonlinear Operators*, Holden-Day, 1964.

[72] M.M. Vainberg, *Variational Method and Method of Monotone Operators in the Theory of Nonlinear Equations*, Wiley, 1973.

[73] G.N. Watson, *A Treatise on the Theory of Bessel Functions*, 2nd edition, Cambridge University Press, 1966.

[74] M. Willem, *Lectures on Critical Point Theory*, Trabalhos de Matemática, University of Brasilia, Brasilia, 1983.

[75] M. Willem, *Minimax Theorems*, Birkhäuser, 1996.

[76] M. Willem, *Periodic oscillations of odd second order Hamiltonian systems*, Boletino U.M.I. 6 (1984), 293–304.

[77] M. Willem, *Density of range of potential operators*, Proc. Amer. Math. Soc. 83 (1981), 341–344.

Index

Printed in the United States of America